Genomic Approaches
for
Cross-Species Extrapolation
in
Toxicology

Other Titles from the Society of Environmental Toxicology and Chemistry (SETAC):

Use of Sediment Quality Guidelines and Related Tools for the Assessment of Contaminated Sediments
Wenning, Batley, Ingersoll, Moore, editors

2005

Ecological Assessment of Aquatic Resources: Linking Science to Decision-Making
Barbour, Norton, Preston, Thornton, editors
2004

Amphibian Decline: An Integrated Analysis of Multiple Stressor Effects
Linder, Krest, Sparling, editors
2003

Metals in Aquatic Systems:
A Review of Exposure, Bioaccumulation, and Toxicity Models
Paquin, Farley, Santore, Kavvadas, Mooney, Winfield, Wu, Di Toro
2003

Silver: Environmental Transport, Fate, Effects, and Models:
Papers from Environmental Toxicology and Chemistry, 1983 to 2002
Gorusch, Kramer, La Point, editors
2003

Contaminated Soils: From Soil–Chemical Interactions to Ecosystem Management
Lanno, editor
2003

Environmental Impacts of Pulp and Paper Waste Streams
Stuthridge, van den Heuvel, Marvin, Slade, Gifford, editors
2003

Porewater Toxicity Testing: Biological, Chemical, and Ecological Considerations
Carr and Nipper, editors
2003

Reevaluation of the State of the Science for Water-Quality Criteria Development
Reiley, Stubblefield, Adams, Di Toro, Erickson, Hodson, Keating Jr, editors
2003

For information about SETAC publications, including SETAC's international journals, *Environmental Toxicology and Chemistry*
Integrated Environmental Assessment and Management, contact the SETAC Administrative Office nearest you:

SETAC Office
1010 North 12th Avenue
Pensacola, FL 32501-3367 USA
T 850 469 1500   F 850 469 9778
E setac@setac.org

SETAC Office
Avenue de la Toison d'Or 67
B-1060 Brussels, Belgium
T 32 2 772 72 81   F 32 2 770 53 86
E setac@setaceu.org

www.setac.org

Environmental Quality Through Science®

Genomic Approaches
for
Cross-Species Extrapolation
in
Toxicology

Edited by
William H. Benson and Richard T. Di Giulio

Proceedings from the Workshop on
Emerging Molecular and Computational Approaches
for Cross-Species Extrapolations

18-22 July 2004
Portland, Oregon USA

Coordinating Editor of SETAC Books
Joseph W. Gorsuch
Gorsuch Environmental Management Services, Inc.
Webster, New York, USA

CRC Press
Taylor & Francis Group
Boca Raton   London   New York

CRC Press is an imprint of the
Taylor & Francis Group, an informa business

Published in collaboration with the Society of Environmental Toxicology and Chemistry (SETAC)
1010 North 12th Avenue, Pensacola, Florida 32501
Telephone: (850) 469-1500 ; Fax: (850) 469-9778; Email: setac@setac.org
Web site: www.setac.org
ISBN 10: 1-880611-88-0 (SETAC Press)
ISBN 13: 978-1-880611-88-3 (SETAC Press)

International Standard Book Number-10: 1-4200-4334-X (Hardcover)
International Standard Book Number-13: 978-1-4200-4334-1 (Hardcover)

Library of Congress Cataloging-in-Publication Data

Genomic approaches for cross-species extrapolation in toxicology / [edited by] William H. Benson,
  Richard T. Di Giulio.
    p. cm.
  Includes bibliographical references.
  ISBN-13: 978-1-4200-4334-1 (alk. paper)
  ISBN-10: 1-4200-4334-X (alk. paper)
  1. Genetic toxicology. 2. Toxicology--Animal models. 3. Toxicity testing.   I. Benson, William H.,
1954- . II. Di Giulio, Richard T. (Richard Thomas), 1950 - .

RA1224.3.G47 2007
616'.042--dc22   2006032674

Visit the Taylor & Francis Web site at
http://www.taylorandfrancis.com

and the CRC Press Web site at
http://www.crcpress.com

and the SETAC Web site at
www.setac.org

SETAC Publications

Books published by the Society of Environmental Toxicology and Chemistry (SETAC) provide in-depth reviews and critical appraisals on scientific subjects relevant to understanding the impacts of chemicals and technology on the environment. The books explore topics reviewed and recommended by the Publications Advisory Council and approved by the SETAC North America, Latin America, or Asia/Pacific Board of Directors; the SETAC Europe Council; or the SETAC World Council for their importance, timeliness, and contribution to multidisciplinary approaches to solving environmental problems. The diversity and breadth of subjects covered in the series reflect the wide range of disciplines encompassed by environmental toxicology, environmental chemistry, and hazard and risk assessment, and life-cycle assessment. SETAC books attempt to present the reader with authoritative coverage of the literature, as well as paradigms, methodologies, and controversies; research needs; and new developments specific to the featured topics. The books are generally peer reviewed for SETAC by acknowledged experts.

SETAC publications, which include Technical Issue Papers (TIPs), workshop summaries, newsletter (*SETAC Globe*), and journals (*Environmental Toxicology and Chemistry* and *Integrated Environmental Assessment and Management*), are useful to environmental scientists in research, research management, chemical manufacturing and regulation, risk assessment, and education, as well as to students considering or preparing for careers in these areas. The publications provide information for keeping abreast of recent developments in familiar subject areas and for rapid introduction to principles and approaches in new subject areas.

SETAC recognizes and thanks the past coordinating editors of SETAC books:

> A.S. Green, International Zinc Association
> Durham, North Carolina, USA
> C.G. Ingersoll, Columbia Environmental Research Center
> US Geological Survey, Columbia, Missouri, USA
> T.W. La Point, Institute of Applied Sciences
> University of North Texas, Denton, Texas, USA
>
> B.T. Walton, US Environmental Protection Agency
> Research Triangle Park, North Carolina, USA
>
> C.H. Ward, Department of Environmental Sciences and Engineering
> Rice University, Houston, Texas, USA

Table of Contents

James S. Bus, Richard A. Canady, Tracy K. Collier, J. William Owens,
Syril D. Pettit, Nathaniel L. Scholz, and Anita C. Street

List of Figures

List of Tables

The Editors

William H. Benson is director of the National Health and Environmental Effects Research Laboratory's Gulf Ecology Division within the US Environmental Protection Agency's Office of Research and Development. Dr. Benson obtained a B.S. degree in biology from the Florida Institute of Technology and his M.S. and Ph.D. degrees in toxicology from the University of Kentucky. While in graduate school, he was the first recipient of the Society of Environmental Toxicology and Chemistry (SETAC) Predoctoral Fellowship sponsored by The Procter & Gamble Company. Dr. Benson has published more than 100 scientific publications focusing on environmental toxicology and chemistry. His research activities have been directed towards assessing the influence of environmental stressors on health and ecological integrity. He has conducted research in the areas of metal and pesticide bioavailability, reproductive and developmental effects in aquatic organisms, endocrine-disrupting chemicals, and use of indicators in assessing health and ecological condition.

Dr. Benson is a past president of SETAC and has served on the International Council of SETAC. He was elected as a fellow of the American Association for the Advancement of Science and is active in several other professional societies. He currently serves as chair of Dow Chemical Company's Technical Advisory Board and as a member of the National Advisory Panel for the Oceans and Human Health Initiative of the National Oceanographic and Atmospheric Administration. In addition, Dr. Benson serves as cochair of the Science Policy Council Coordinating Committee for Development of a Technical Framework and Training for Genomics within the US Environmental Protection Agency (USEPA) and member of the steering team for the ILSI Health and Environmental Sciences Institute Committee on the Application of Genomics to Mechanism-Based Risk Assessment.

Richard Di Giulio is professor of environmental toxicology in the Nicholas School of the Environment and Earth Sciences at Duke University, Durham, North Carolina. At Duke, he also serves as director of the Integrated Toxicology Program (a doctoral and postdoctoral training program), director of the Superfund Basic Research Center, and director of the Center for Comparative Biology of Vulnerable Populations (all supported principally by the National Institute of Environmental Health Sciences (NIEHS)). He received a B.A. in comparative literature from the University of Texas at Austin, an M.S. in wildlife biology from Louisiana State University, and his Ph.D. in environmental toxicology from Virginia Polytechnic Institute and State University.

Dr. Di Giulio has published more than 100 scientific research papers on subjects including biochemical and molecular mechanisms of adaptation and toxicity, xenobiotic metabolism and oxidative stress, genotoxicity and gene expression, and developmental toxicity. Most of this work has employed aquatic organisms. Additionally, he has organized symposia and workshops and written on the broader subject of interconnections between human health and ecological integrity. Dr. Di Giulio serves as an advisor for the Scientific Advisory Board of the USEPA and is a member of the Computational Toxicology Committee for the Board of Scientific Counselors, USEPA, and of the National Academy of Science Committee on Assessment of the Health Implications of Exposure to Dioxin. He has served on the board of directors for the Society of Environmental Toxicology and Chemistry, is active in the Society of Toxicology, and is associate editor for *Toxicological Sciences*.

Contributors

James S. Bus
The Dow Chemical Company
Midland, Michigan, USA

Richard A. Canady
US Food and Drug Administration
Rockville, Maryland, USA

John K. Colbourne
Indiana University
Bloomington, Indiana, USA

Tracy K. Collier
National Oceanic and Atmospheric
  Administration
Seattle, Washington, USA

Jon C. Cook
Pfizer
Groton, Connecticut, USA

Nancy D. Denslow
University of Florida
Gainesville, Florida, USA

David Dix
US Environmental Protection Agency
Research Triangle Park, North Carolina,
  USA

David L. Eaton
University of Washington
Seattle, Washington, USA

Jonathan H. Freedman
Duke University
Durham, North Carolina, USA

Evan Gallagher
Univeristy of Washington
Seattle, Washington, USA

Caren C. Helbing
University of Victoria
Victoria, British Columbia, Canada

Michael J. Hooper
Texas Tech University
Lubbock, Texas, USA

Taisen Iguchi
Okazaki National Research Institutes
Okazaki, Japan

Sean Kennedy
Environment Canada
Ottawa, Ontario, Canada

Elwood A. Linney
Duke University Medical Center
Durham, North Carolina, USA

Renae L. Malek
The Institute for Genomic Research
Rockville, Maryland, USA

Ann Miracle
US Environmental Protection Agency
Cincinnati, Ohio, USA

J. William Owens
The Procter and Gamble Company
Cincinnati, Ohio, USA

Syril D. Pettit
ILSI Health and Environmental
  Sciences Institute
Washington, DC, USA

John Quakenbush
The Institute for Genomic Research
Rockville, Maryland, USA

Kenneth S. Ramos
University of Louisville
Louisville, Kentucky, USA

Dan Schlenk
University of California, Riverside
Riverside, California, USA

Patricia K. Schmeider
US Environmental Protection Agency
Duluth, Minnesota, USA

Nathaniel L. Scholz
National Oceanic and Atmospheric
  Administration
Seattle, Washington, USA

Joseph R. Shaw
Dartmouth College
Hanover, New Hampshire, USA

Ilya Shmulevich
The University of Texas
Houston, Texas, USA

Anita C. Street
US Environmental Protection Agency
Washington, DC, USA

Joshua M. Stuart
University of California–Santa Cruz
Santa Cruz, California, USA

Claudia Thompson
National Institute of Environmental
  Health Sciences
Research Triangle Park, North Carolina,
  USA

Mark R. Viant
University of Birmingham
Birmingham, England, UK

Michael Waters
US Environmental Protection Agency
Durham, North Carolina, USA

Phillip L. Williams
University of Georgia
Athens, Georgia, USA

Timothy R. Zacharewski
Michigan State University
East Lansing, Michigan, USA

Preface

Advances in molecular technology have led to the elucidation of full genomic sequences of several multicellular organisms, ranging from nematodes to man. The related molecular fields of proteomics and metabolomics are now beginning to advance rapidly as well. In addition, advances in bioinformatics and mathematical modeling provide powerful approaches for elucidating patterns of biological response embedded in the massive datasets produced during genomics research. Thus, changes or differences in the expression patterns of entire genomes at the levels of mRNA, protein, and metabolism can be assessed rapidly. Collectively, these emerging approaches may greatly enhance our ability to address many of the major issues in human and environmental toxicology. Specifically, they are uniquely qualified to address the issue of cross-species extrapolation in risk assessment in human and environmental toxicology.

Although there may be important differences in genomes and proteomes among species, many of the responses to various stressors are evolutionarily conserved. For example, consider how fish, birds, and mammalian species respond to external stressors, including chemical toxicants (synthetic and natural), genotoxicants (carcinogenic or mutagenic), or parasites. Stressed organisms can initiate defensive and offensive actions to counteract adverse responses. Many of these defensive responses to external stimuli are common to many organisms, including wildlife species (fish, birds, invertebrates) and humans. Genomic technologies may provide great insight into how diverse organisms respond to environmental stressors.

Motivated by these concerns, the Society of Environmental Toxicology and Chemistry (SETAC) and the Society of Toxicology (SOT) jointly sponsored a workshop entitled "Emerging Molecular and Computational Approaches for Cross-Species Extrapolations" in Forest Grove, Oregon, from July 19 to July 22, 2004. This workshop was significant because leading societies — concerned on the one hand with the integrity of the environment (SETAC) and on the other hand with the improvement of human health (SOT) — worked together. Thirty-five scientists and professionals were brought together from diverse fields, including environmental toxicology and chemistry, biomedical toxicology, molecular biology, genetics, physiology, bioinformatics, computer science, and statistics. Such collaboration provided an ideal vehicle for objective and balanced discussion of this topic among professionals from different yet highly interrelated disciplines.

In addition, a 2002 workshop on human health and ecological integrity (Di Giulio and Benson 2002) provided a basis for this workshop on computational approaches and cross-species extrapolations. The overall goal of the workshop was to outline a research agenda utilizing emerging technologies in omics and computational biology in order to 1) elucidate similarities and differences among species, 2) relate stressor-mediated responses to adverse outcomes, and 3) extend this science into innovative approaches to risk assessment and regulatory decision-making.

Workshop participants identified specific research gaps and emerging issues and key conclusions. Key conclusions and recommendations from the workshop were as follows:

- Improve genomic technologies that currently provide powerful research tools but are insufficient as a basis for risk assessment and replacement of traditional approaches.
- Perform collaborative proof-of-concept studies to improve our understanding of cross-species extrapolation by characterizing similarities and differences in metabolic pathways.
- Develop more standardized approaches for omics technologies and associated data analysis.
- Develop genomic databases for selected surrogate species focusing on basic, conserved cellular and physiologic processes.
- Perform studies to validate the relationship between omics responses and adverse biological outcomes.
- Form a standing task force for cross-species and genomic issues.
- Enhance training in genomic technologies, particularly within the context of an interdisciplinary approach.

This book presents all relevant discussions and conclusions from the 2004 workshop. The workshop organizers and participants hope this book provides a meaningful starting point for future discussions regarding the use of emerging molecular and computational approaches for cross-species extrapolations.

William H. Benson
Richard T. Di Giulio

REFERENCE

Di Giulio RT, Benson WH, editors. 2002. Interconnections between human health and ecological integrity. Pensacola (FL): SETAC Press.

Acknowledgments

The SETAC–SOT workshop was sponsored by

- National Institute of Environmental Health Sciences
- National Oceanographic and Atmospheric Administration
- Pfizer, Inc.
- The Procter & Gamble Company
- US Environmental Protection Agency

1 "Omics" Approaches in the Context of Environmental Toxicology

Jon C. Cook, Nancy D. Denslow, Taisen Iguchi, Elwood A. Linney, Ann Miracle, Joseph R. Shaw, Mark R. Viant, and Timothy R. Zacharewski

1.1 INTRODUCTION

The goals of this chapter are to 1) introduce genomics, transcriptomics, proteomics, and metabolomics technologies; 2) describe the advantages and challenges associated with these approaches compared to traditional methodologies, particularly from the perspective of cross-species extrapolation within human and environmental toxicology; and 3) identify solutions that will facilitate the incorporation of these technologies into environmental toxicology.

1.2 OVERVIEW OF OMICS TECHNOLOGIES

The ability to measure hundreds or thousands of genes, proteins, or metabolites from a single sample has been commonly referred to by the suffix "omics." The 4 general categories of omics technologies include genomics, transcriptomics, proteomics, and metabolomics (Table 1.1). Before the invention of these technologies, genes were studied using Southern blot analysis or gene sequencing, transcripts were studied using northern blot analysis or RNase protection, proteins were studied using western blot analysis or enzyme-linked immunosorbent assay (ELISA), and metabolite concentrations were typically measured using enzymatic assays or chromatographic approaches.

The term "genomics" is often used interchangeably with transcriptomics to refer to gene-expression studies and occasionally is used to describe all omics approaches. However, for the current discussion, genomics will refer only to genome-level (i.e., DNA sequence) omics applications, an area of study that includes population genetics, haplotype analysis, genomics, quantitative trait locus (QTL) and single nucleotide polymorphism (SNP) expression analyses, and resequencing (Hacia 1999; Jansen and Nap 2001; Luikart et al. 2003; Sen and Ferdig 2004). Genomics is the functional

TABLE 1.1
Four general categories of omics technologies

Technology	Type of sample	Methodology	Traditional research equivalent
Genomics: assesses entire genome	DNA	DNA–microarray; QTL and SNP maps; resequencing	Gene sequencing, Southern
Transcriptomics: assesses functional genome	RNA	DNA–microarray	Northern, RNase protection
Proteomics	Protein	Two-dimensional gel electrophoresis; ELISA–microarray	Western, ELISA
Metabolomics	Metabolites (in tissues or body fluids)	NMR spectroscopy; mass spectrometry	Enzymatic assay; HPLC (UV detector)

HPLC = high performance liquid chromatography.

Source: Modified from Martin R, Leppert D. 2005. J. Neurological Sci. 222, 3–5.

study of genetic variation within a population, identifying genome of sequence variability (i.e., polymorphisms) associated with a phenotype or response of interest. It provides a functional link between genotype and individual variability within a population. The main tools for the study of genomics are the DNA microarray (commonly referred to as "DNA chip") and other genome scanning tools used to identify polymorphisms (e.g., QTL and SNP maps; resequencing).

Transcriptomics is the study of the full complement of activated genes that encode mRNAs or transcripts in a tissue or sample. Proteomics is defined as the study of the full set of proteins encoded by the genome, although this is not currently possible due to technical limitations. The protein component has been estimated to be at least 1 order of magnitude larger than the genome and includes proteins that result from alternative splicing and post-translational modifications. Proteomics-based microchip platforms are at a much earlier stage of development than DNA microarrays. Proteomics uses a combination of techniques, including gel electrophoresis, chromatographic or laser-based separation techniques coupled with mass spectrometry or antibody-based detection approaches.

The complete characterization of the functional phenotypic responses of a cell, tissue, or organism to a drug or toxicant also necessitates the measurement of low molecular weight metabolites, which form the substrates of a plethora of enzyme-mediated biochemical processes. The comprehensive characterization of cellular metabolite concentrations is termed "metabolomics" and is the newest of the omics technologies. Metabolomics typically employs techniques such as nuclear magnetic resonance (NMR) spectroscopy or mass spectrometry. The interrelationships among genomics, transcriptomics, proteomics, and metabolomics are illustrated in Figure 1.1.

In order to fully assess the potential adverse health effects of chronic and subchronic exposure to synthetic and natural chemicals and their complex mixtures

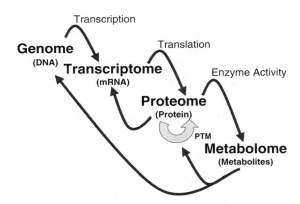

FIGURE 1.1 The interrelationships among genomics, transcriptomics, proteomics, and metabolomics. PTM = post-translational modification.

in the environment, a more comprehensive understanding of the molecular-, cellular-, and tissue-level effects of these compounds is required within the context of the whole organism. The availability of genomics sequence for a number of toxicologically relevant species and recent advances in global assessment technologies that can simultaneously measure changes in the levels of many gene transcripts, proteins, and metabolites have provided unprecedented opportunities to elucidate mechanisms of toxicity, identify susceptible populations and novel biomarkers, and support human and ecological risk assessment.

Traditional toxicological studies have generally been limited to examining hypotheses focused on a preselected single response (i.e., change in expression of a single gene or in a single enzyme activity) that is temporally or dose-dependently monitored following exposure to a toxicant. In contrast, omics technologies allow investigators to examine, in an unbiased manner, comprehensive changes in genomics sequence, transcript, protein, and metabolite levels. This approach, in principle, provides the opportunity to identify previously undiscovered pathways perturbed by toxicant exposure and interactions between pathways that provide a more comprehensive assessment of molecular and biochemical effects of exposure. Comparative approaches further assess toxicity by identifying conserved responses among species and extend the elucidation of modes of action by interrogating complementary targets within orthologous pathways. Moreover, conserved pathways in comparative species analyses may be useful in other technologies (i.e., antisense morpholino knockdown in zebrafish) that can further probe the networks and mechanisms of toxicity.

The incorporation of omics approaches also has the potential to identify more predictive biomarkers. In general, the applicability of biomarkers to support environmental risk assessment has been hindered by 1) lack of specificity towards a particular stressor (e.g., class of pollutants), 2) variability in biomarker responses due to seasonal changes in organism biochemistry and the combined effects of multiple stressors, and 3) insufficient evidence that suborganismal changes will have a significant impact on organism or population health. These limitations arise due to the reliance on a single response that is multifactorally regulated and therefore subject to false positive and negative rates. However, comprehensive omics technologies and associated bioinformatics methods may provide the opportunity to define specific profiles for stressors that will be more discriminating and less subject to other influences, thus enhancing the overall predictivity as well as allowing application of a systems biology approach to environmental toxicology.

1.3 DISCOVERY-DRIVEN VERSUS HYPOTHESIS-DRIVEN RESEARCH: A NEED FOR BALANCE

Omics technologies have provided powerful approaches as well as some significant challenges (Table 1.2). The primary advantage of the omics approaches is their ability to provide comprehensive support of "hypothesis-driven" and "discovery-driven" research. Traditional hypothesis-driven research has focused on 1 gene, protein or enzymatic activity, and, in many cases, outside of relevant biological context, in order to explain complex toxicological responses. In this targeted approach, a hypothesis is conceived on the basis of previous knowledge of the system under investigation, and then a focused dataset is collected that specifically tests the hypothesis. Although these studies have provided valuable points of reference, emerging omics technologies provide the opportunity to significantly advance the elucidation of mechanisms responsible for disease states and toxicity by examining multiple biological processes simultaneously. For example, gene expression data can be used to address specific questions, such as whether a compound is simulating an estrogen receptor agonist, by comparing the response of the array with that obtained with a positive control like 17β-estradiol.

In contrast, discovery-driven research is hypothesis free and can capture information that extends current knowledge regarding a pathway as well as interactions between pathways. This approach provides unbiased analysis of a large number of targets simultaneously, using genomics, transcriptomics, proteomics, or metabolomics. Although it is extremely powerful, hypothesis-free research funding by federal agencies has been rare compared to that for hypothesis-driven research. There are clear examples where discovery-driven research needs to be funded to advance scientific understanding. For example, a case has been made for the application of discovery-driven omics approaches to study multiple sclerosis because 30 years of hypothesis-driven research has failed to identify an effective treatment for this disease (Martin and Leppert 2004).

Discovery- and hypothesis-driven approaches are complementary; they can further elucidate specific pathways involved in a toxic response as well as place the pathway into biological context with other elicited effects. Clearly, discovery-driven research can play a valuable role in biomedical research by rapidly identifying in an unbiased manner several hypotheses for treating uncharacterized diseases. Fortunately, the importance of hypothesis- and discovery-driven research in environmental toxicology is gaining acceptance and is a primary goal of the National Center for Toxicogenomics' Chemical Effects in Biological Systems (CEBS) knowledge base (Waters et al. 2003).

TABLE 1.2
Overview of advantages, challenges, and solutions of omics technologies

Technologies	Advantages	Challenges	Solutions
Genomics (Human: 1.42 million SNPs database compared to estimated 11 million)	1) Phenotype distribution within population 2) Link phenotype to genotype 3) Susceptibility and haplotype map	1) Full understanding of genetics is necessary to utilize QTLs 2) Tractable genetic architecture 3) SNPs require large scale sequencing	1) More sequencing of species and of individuals within a population
Transcriptomics (Human: 30,000 mRNAs)	1) Most well developed of omics technologies 2) Off-the-shelf availability 3) Most complete coverage (selected organism) 4) Most annotated	1) Limited number of species 2) Relevance to phenotype 3) Across-platform inconsistencies	1) Availability of species-specific arrays 2) Phenotypic anchoring 3) Standardization
Proteomics (Human: 300,000 proteins)	1) Closer to contributing to phenotype than transcriptomics 2) Identification of gene products 3) Identification of post-translational modifications (e.g., phosphorylation/ glycosylation)	1) Detection limit of low-level proteins 2) Large number of proteins–isoforms 3) Quantification	1) Methods development for fractionation of proteins 2) Development and use of non-gel-based methods 3) Development of specific antibodies for ELISAs, western blots and isotopic labeling
Metabolomics (Human: 2000–20,000 metabolites)	1) Defines the molecular phenotype 2) No genome required, which facilitates cross-species extrapolation 3) Integrates genetics and environment 4) Inexpensive per-sample analysis cost 5) Most reproducible of omics technologies 6) Potentially quantitative	1) Peak identification (NMR and MS) 2) Extraction bias for tissue samples (NMR and MS) 3) Detection limit for low-concentration metabolites (NMR)	1) Construction of metabolite databases 2) Development of extraction methods 3) Development of higher sensitivity NMR methods and use of other technologies

MS = mass spectroscopy.

1.4 ADVANTAGES, CHALLENGES, AND SOLUTIONS OF OMICS TECHNOLOGIES

Table 1.2 summarizes the key advantages, challenges, and solutions of 4 omics technologies. Each of these technologies is discussed in the following sections, and key examples are provided. There are several generic challenges impeding the widespread incorporation of all omics technologies into environmental toxicology, including cost (infrastructure, operating expenses, consumables); limited expertise; lack of consensus regarding data processing, analysis, and management; complexity of data interpretation, knowledge extraction, and pathway elucidation; and incorporation of omics data into risk assessment. However, there has been significant progress since the introduction of these technologies, and the likelihood is that impediments will be overcome as the limitations are further defined and these approaches are applied with greater frequency to address complex environmental problems.

1.4.1 ADVANTAGES OF GENOMICS APPROACHES

In its broadest sense, genomics is the genome-wide study of DNA sequence information. Given that the genetic code defines the heritable attributes of individuals, sequencing of genomes offers insights into different individual responses to environmental agents and associations between genotype and phenotype. From a toxicological perspective, phenotypic variability describes the conditional responses of individuals exposed to a toxicant. In fact, such phenotypic variability is the basis for dose–response relationships, which characterize differences among individuals in their susceptibilities to toxicants to predict population responses (Eaton and Klaassen 2001).

A measurable portion of phenotypic differences is characterized by DNA sequence variability (i.e., polymorphisms, haplotypes). Polymorphisms within and among species or populations exist as SNPs, insertion and deletion (indels) mutations, or repeated motifs. While the overall frequency of polymorphisms is usually low among populations, SNPs are the most prevalent variants and are the major genetic contributor to phenotypic variability. In fact, their occurrence at a frequency >1% is typically used to define and distinguish them from germline mutations that occur less frequently (<1%).

The environment also contributes to phenotype, complicating the task of functionally characterizing toxicologically relevant polymorphisms. However, genomics provides the tools needed to address some of these challenges. Unlike other omics technologies, it is an integrated process that applies several resources. Genomics compares the physical (i.e., genome sequence) and genetic maps, across individuals and populations, to explore the genotypic basis underlying the origin, stability, and distribution of relevant phenotypes in a population. Collectively, genomics resources cluster into 2 areas based on their applications: 1) characterization of quantitative genetic variation and 2) polymorphism-scanning or mapping to identify the molecular underpinnings of the variation (Farrall 2004; Sen and Ferdig 2004). These technologies include

- population genetic techniques such as QTL analyses to link phenotype (quantitative trait) within regions (loci) of the genome (Lynch and Walsh 1997),
- primary sequence information to identify genes and other functional units within this region,
- the genetic map to characterize the recombinational architecture within this region,
- comparative sequence analysis of genes within a population (i.e., DNA microarrays, resequencing), and
- reverse-transcription PCR (polymerase chain reaction) to identify splice variants and alternative regulatory regions of genes.

While toxicological applications of these integrated genomics technologies have been suggested and are in current use, they have yet to appear in the literature. However, their promise to toxicology is highlighted by examples that demonstrate their utility in identifying genotypes associated with disease susceptibility and applicability to cross-species extrapolation. QTLs represent physical regions of the genome that contain polymorphisms affecting a phenotype of interest. "Mapping" refers to the process of identifying these genomics regions and characterizing their relative effects and mechanisms of action on the phenotype. Mapping provides insight into the intricacy of genetic control (e.g., single- or multiloci) because the distribution of phenotypic variance is indicative of the genetic complexity. For example, a quantitative trait will distribute continuously in a population, which suggests that numerous independent loci work discretely (i.e., polygenic, epistasis) to influence phenotype.

Whereas a single-gene effect (e.g., Mendelian) segregates into 2 classes (bimodal), many common disease factors (e.g., height, weight, hypertension) and toxicologically relevant phenotypes are thought to be polygenic. This complexity represents a major challenge to the field of toxicology. One method used to identify loci employs genetic crosses between pairs of individuals with divergent phenotypes. Depending on the effect of loci (additive, multiplicative, epistatic), their progeny would display a range of phenotypic outcomes with diagnostic genetic signatures (i.e., genotype) because of meiotic recombination. Because each genotype encodes a specified phenotype, these signatures can be used to map traits by finding markers that cosegregate with the phenotype.

Genomics markers (e.g., haplotypes, SNPs, microsatellite markers) are used to track rearrangements in the genomics architecture and to localize chromosomal regions containing genes with significant phenotypic responses (e.g., toxic response). The size of a mapped locus is, in part, determined by chromosome position because less recombination occurs at the centers versus the ends of chromosomes. Map distances can be further reduced through additional crossing until a point is reached where recombination and saturating the genome with tightly linked markers cannot randomly distribute the phenotype (i.e., linkage disequilibrium; Kirk et al. 2002).

Once regions of the genome containing polymorphisms associated with the phenotype of interest have been identified, they can be overlaid with sequence data, high-density SNP mapping, or gene expression studies to identify the molecular basis of phenotype variation. This approach was employed by Hoekenga et al. (2003) to investigate aluminum tolerance in *Arabadopsis thaliana*. QTL analysis identified 2 loci associated with aluminum tolerance. These regions each spanned ~3 Mb and contained ~700 predicted genes. DNA microarrays revealed that 15 of the genes within these 2 regions were responsive to aluminum, and a polymorphism map indicated that 13 of these contained SNPs in their coding or noncoding regulatory regions.

Genotype variation in regulatory regions of the genome is perhaps of greatest importance for cross-species interpretations because of the generally higher level of sequence conservation in coding regions of genes. In fact, King and Wilson (1975) hypothesized that variation in gene expression (i.e., noncoding region polymor-phisms and to a lesser extent polymorphisms in transcription factors) accounts for the dramatic differences in phenotype observed between closely related species (i.e., humans and chimpanzees). This hypothesis is gaining experimental support from microarray experiments in fish (Oleksiak et al. 2002), primates, and mice (Enard et al. 2002) and suggests that comparisons of orthologous gene expression could provide a link between phenotypic or response differences observed between species. Furthermore, genomics as described in this chapter has the potential to improve such strategies for cross-species extrapolations.

Using a population-based genomics approach, comparisons of orthologous genes could focus not only on those associated with key pathways, but also on those found in regions of the genome that associate with a phenotype of interest (e.g., orthologous QTLs) or, if identified, the underlying polymorphisms. Furthermore, if the syntenic relationship between the species being compared were known, it could highlight comparable genomics regions based on the genetic map. Vitt and coworkers (2004) used a similar approach to study disease phenotypes in rats and humans. These authors identified a region of rat chromosome 1 that contained several QTLs related to disease (i.e., renal disease, diabetes, hypertension, body weight, encephalomyeli-tis). They aligned and clustered ~1.5 million rat expressed sequence tags (ESTs) to the rat genome to identify genes that fell within the identified coordinates of the chromosome. The human–rat syntenic map revealed rearrangement from chromo-some 1 of the rat to chromosomes 9 and 10 of the human.

Syntany mapping also provided support for functional identity of the rat genes because of sequence similarities between rat and human genes as well as the con-sistent location of the genes on the genome. Disease linkage data revealed a common QTL — the rat renal failure-1 (Rf-1) region and the human orthologue, end-stage renal disease (ESRD) — within the identified areas of rat and human chromosomes, respectively. With the additional information provided from the human ESRD region, the size of the rat QTL was reduced by half (i.e., from 20 to 11.5 Mb). This information was combined with data from expression studies to identify a set of 66 Rf-1-related candidate genes in the rat kidney.

1.4.2 CHALLENGES OF GENOMICS APPROACHES

While QTL analyses, SNP mapping, expression-coupled QTL analyses, and other genomics approaches offer great promise to toxicology, there are considerable challenges associated with their application. At the forefront of these challenges is the limited number of organisms that have defined annotated genomes and for which the genetic tools needed to map traits to the genome exist. In addition, the number of organisms that have considerable resequencing efforts is even smaller. Thus, for toxicology studies that utilize one of the current model organisms, such applications hold promise. For studies that employ nonmodel organisms, considerable resources need to be developed for these approaches to be applicable. This issue is perhaps the most germane to cross-species extrapolations because it is difficult to extrapolate to the unknown. Another challenge associated with genomics applications lies in the anticipated complexity (e.g., polygenic, epitasis, gene environment) of toxicologically relevant phenotypes. These challenges are compounded by the fact that the environment also contributes to phenotype.

While pharmaceutical and biomedical research has employed genomics approaches to identify candidate genes or regulatory elements associated with a given phenotype, confirmatory studies are lacking. For example, it is known that polymorphisms in the family of phase II metabolizing enzymes, glutathione S-transferases (GSTs), influence their activity (Daly 2003) and are associated with clinical outcomes in cancer patients (Innocenti and Ratain 2002). Nevertheless, there is little consensus on their utility in determining disease risks, and direct correlations between genotype and phenotype are lacking. It has been 5 years since the announcement of the initial human genome sequence, which was hailed as a promise of improved health care and drug therapies through personalized medicine. Some even further suggested that the potential for personalized risk assessment customized to an individual genotype was possible in the near future (Thomas et al. 2002). Despite considerable fanfare, these promises have yet to be achieved, and debate continues on the magnitude of their future potential.

1.4.3 SOLUTIONS OFFERED FOR GENOMICS APPROACHES

As with most technological challenges, the solution lies with time and resources. Obviously, increasing the number of annotated genome sequences will improve genomics comparisons, especially those between species. Understanding the genetic and environmental contributions to complex polygenic phenotypes will require coupling genomics applications with technologies that probe downstream processes (i.e., transcriptome, proteome, metabolome). Such an integrative, "systems toxicology approach" will also require considerable investment in computational resources (e.g., statistical approaches, bioinformatics platforms, mathematical modeling). The development of resources that accurately integrate "global" molecular datasets generated from omics technologies and facilitate biological understanding across levels of organization and species boundaries is daunting and will require "new methods of data management, data integration, and computational modeling" (Waters and Fostel 2004).

1.4.4 VALIDATION OF GENOMICS

An understanding of the genetic bases of phenotype variation is a powerful tool that offers promise for identifying gene networks underpinning complex phenotypes and defining susceptibility to exposure. However, it must be stressed that it does not define molecular mechanisms that occur downstream of the genome (i.e., transcriptome, proteome, metabolome). For example, the genetic basis of cystic fibrosis is due to mutations in a single gene: the cystic fibrosis transmembrane conductance regulator (CFTR) cloned in 1989 (Riordan et al. 1989). Despite its being one of the most intensely studied proteins in biomedical sciences, this information has not resulted in a cure for the disease (e.g., the median age of survival for a person with cystic fibrosis is in the mid-30s), largely because the functional mechanisms downstream of the genome (e.g., transcriptome, proteome, metabolome) that regulate phenotypic expression have yet to be fully identified.

Peakall and Shugart (2002) noted the importance of a comprehensive approach in defining the molecular events associated with the toxic state and establishing "causation between pollution exposure and effects at all levels of biological organization." This concept has been referred to as "phenotype anchoring" and has been cited as a critical need for toxicology and risk assessment omics applications (see Chapter 5 that discusses risk assessment and regulatory decision making). Such a comprehensive approach, although critical, lacks the basic experimental support that is a necessary step to validating these and other omics technologies.

1.4.5 POTENTIAL OF GENOMICS APPROACHES FOR ECOTOXICOLOGY

The application of integrated genomics approaches as described previously, to the field of ecotoxicology will be difficult because they require resources (e.g., annotated genome, genetic map, individual sequence data) available only for a limited number of model organisms. Of these, the primary limitation is the absence of a sequenced genome for any of the prominent ecotoxicological species. This limits current genomics applications to comparative studies between model organisms and ecotoxicologically critical models. Using this paradigm, the zebrafish could be used to identify toxicologically relevant polymorphisms and the network of genes they influence; these could then be cloned and studied in the fathead minnow. This approach is strengthened with the development of other omics resources for fathead minnow (e.g., DNA microarrays). Future promise lies in the development of ecotoxicological critical models. In this regard, *Daphnia pulex* is probably the most advanced. Its genome is being sequenced, and efforts are underway to develop complementary resources (e.g., genetic map, microarrays, cell lines, expression vectors, microsatellite markers, genomics and cDNA libraries, bioinformatics platform) to increase the utility of the annotated genomics map.

1.4.6 TRANSCRIPTOMICS

Of the omics technologies, microarrays are the most well established and provide the greatest breadth; some commercial arrays provide whole genome coverage for selected toxicologically relevant mammalian species such as human and mouse. Extensive annotation of the genes represented on these commercial arrays continues to increase and is another significant advantage that facilitates the conversion of microarray data into mechanistic and pathway knowledge to support risk assessment.

However, full utility of microarray technology to risk assessment is limited by a number of factors, including the number of different platforms (e.g., cDNA arrays, oligonucleotide arrays, Affymetrix GeneChips) and multicenter cross-validation studies that indicate significant inconsistencies in data between laboratories. Although differences in microarray technologies are significant contributing factors, the lack of standardized protocols between laboratories and differences in data processing and analysis also significantly contribute to variability in assay reproducibility and sensitivity.

Some of these differences may be accommodated within analyses, a requirement of the emerging minimum information about a microarray experiment (MIAME) standards (Brazma et al. 2001). However, the likelihood of consensus protocols and compliance in the user community is questionable. The relevance of changes in transcript levels to toxicity requires verification by independent methods such as quantitative real-time polymerase chain reactin (PCR), in situ hybridization, western analysis, immunohistochemistry, or functional analyses (Rockett and Hellmann 2004). These latter approaches further validate the significance of the change in gene expression and localize gene expression change within multicellular tissues; this could provide valuable complementary data or "phenotypic anchoring" in network and pathway elucidation.

Measuring changes in gene expression over time and across dose provides critical information regarding the kinetics and coordination of gene expression that contributes to the dynamic processes of cellular homeostasis and toxicity. Potential confounding secondary to toxicity can be minimized by selecting and studying the genes mediated by the pharmacologic response of interest; selection of these genes is dependent upon the availability of positive (agonist and antagonist) and negative controls (Koza-Taylor et al. 2005). Analyzing gene expression data across multiple samples or responsive species also reveals underlying similarities among different conditions, thus producing correlates of gene behavior that can be used to predict and diagnose cellular responses to exogenous chemicals.

Microarray experiments have been applied to tumor samples to demonstrate the potential of global gene expression profiling to accurately diagnose disease phenotypes based on gene expression alone (Alon et al. 1999; Perou et al. 1999; Alizadeh et al. 2000). It is anticipated that similar profiling strategies can be used to investigate the modes of action of chemicals and to classify them based on the similarity of their expression profiles compared to expression profiles obtained from known chemicals with defined mechanisms of action. Proof-of-principle experiments have demonstrated this in a variety of model systems, including yeast (Marton et al. 1998; Hughes et al. 2000), cultured mammalian cells (Burczynski et al. 2000; Waring, Cuirlionis et al. 2001), and rodent liver (Thomas et al. 2001; Waring et al. 2001; Hamadeh, Bushel, Jayadev, Di Sorbo, et al. 2002; Hamadeh, Bushel, Jayadev, Martin, et al. 2002).

The development of transcriptomics during the past 10 years has focused on human health research using mammalian species or developmental biology involving *Caenorhabditis elegans* and *Drosophila*. Its advantages (Table 1.2) have only recently been realized for a few ecotoxicology animal models that have been characterized by full genome or EST annotation (e.g., medaka, *Xenopus laevis*, zebrafish). Although not widely available, several independent laboratories have produced low-density, cDNA, or EST arrays for a number of other ecotoxicology relevant species (e.g., European flounder, sheepshead minnow, largemouth bass, and *Xenopus laevis*) (Table 1.3) with ecotoxicogenomics as the focus for tool development (Larkin et al. 2003; Williams et al. 2003; Knoebl et al. 2004; Sone et al. 2004). While these latter efforts contribute to the overall field, they do not reflect breadth made possible with this technology; hence, ecotoxicology stands to make significant advances provided the full benefits are realized.

TABLE 1.3
Current status of transcriptomics tools available for ecotoxicology application

Species	Array format	Number of features	Resource
Chicken	Oligo	42,000	http://www.affymetrix.com/products/
	cDNA	13,000	genomics@fhcrc.org
Zebra finch	cDNA	2500	http://titan.biotec.uiuc.edu/songbird/
Alligator	EST	10,000	taisen@nibb.ac.jp
Goldfish	cDNA	400	trudeauv@uottawa.ca
Xenopus laevis	Oligo	14,400	http://www.affymetrix.com/products/
	EST	40,000	ueno@nibb.ac.jp;
Rana sp. and *Xenopus* sp.	cDNA	400	http://www.viagenx.ca
Medaka	Oligo	10,000	Nagakama — NIBB
	Oligo	8000	htakeda@biol.s.u-tokyo.ac.jp
Zebrafish	Oligo	14,900	http://www.affymetrix.com/products/
	Oligo	14,000	http://www.mwg-biotech.com/
	Oligo	22,000	http://www.chem.agilent.com/
Fathead minnow	Oligo[a]	2500	http://www.ecoarray.com/
Largemouth bass	cDNA	500	http://www.ecoarray.com/
Sheepshead minnow	cDNA	250	http://www.ecoarray.com/
European flounder	cDNA	12,000	http://www.genipol.stir.ac.uk/
Rainbow trout	cDNA	200	mvijayan@sciborg.uwaterloo.ca
Salmon (GRASP)	cDNA	16,000	http://web.uvic.ca/cbr/grasp/
Fundulus heteroclitus	cDNA	10,000	http://crawford.rsmas.miami.edu/
Tomcod	cDNA	6000	wirgin@env.med.nyu.edu
Catfish	cDNA	660	zliu@acesag.auburn.edu
Ciona intestinalis	Oligo	22,000	kaoru@pharm.hokudai.ac.jp.
Daphnia magna	EST	10,000	taisen@nibb.ac.jp
Daphnia pulex	cDNA	2500	http://daphnia.cgb.indiana.edu/tools/microarrays/
Mysid shrimp	cDNA	100	marius.brouwer@usm.edu
Mytilus californianus	cDNA	4224	paola.venier@unipd.it
Drosophila melanogaster	Oligo	18,500	http://www.affymetrix.com/products/
Honey bee	ESTs	9000	grozinge@staff.uiuc.edu
Anopheles gambiae	ESTs	3840	kafatos@embl-heidelberg.de
Caenorhabditis elegans	Oligo	22,490	elegans.array@watson.wustl.edu
	Oligo	22,500	http://www.affymetrix.com/products/

[a] Not released, but will be based on an Agilent or Amersham platform.

Table 1.3 provides a list of some ecotoxicology and potential surrogate model organisms and the availability of transcriptomics tools. The solutions for overcoming the challenges facing ecotoxicogenomics are simple in part, but require significant resources. As more ecotoxicology model organisms gain genome or EST information, the availability of species-specific arrays for a variety of sentinel species for ecotoxicology should increase.

In the interim, the application of transcriptomics in ecotoxicology may rely on the use of organisms with full transcriptome or EST-derived tools as surrogates for relevant ecotoxicology models (e.g., use zebrafish and zebrafish arrays to investigate stressor responses in the fathead minnow). The mining of gene expression changes in a surrogate can provide information regarding potentially conserved pathways and biological processes that are affected by specific stressors. Validation of the surrogate response can be assessed in ecotoxicology models in the field by using "caging-type study designs" and assessing phenotypic endpoints or a variety of molecular techniques that require less genome information. The potential of being able to link transcriptional changes with measurable effects supports the use of ecotoxicology models by taking advantage of existing animal toxicology data generated in a number of established organisms commonly used in regulatory studies (USEPA 1998; OECD 2004).

Challenges similar to those in mammalian toxicology face the use of transcriptomics in ecotoxicology. Without linkage to mechanistic pathways or relevant endpoints, gene expression data lack the proof of concept that a given stressor significantly contributes to a given adverse effect. The lack of ecotoxicologically relevant, commercially available arrays also leads to complications with platform inconsistencies. With independent efforts producing perhaps redundant yet different products, the ability to compare data across spotted cDNA, EST, or oligonucleotide arrays becomes compromised. As mentioned previously, standardization efforts for reporting data from transcriptomics experiments have been greatly aided by the efforts of the Microarray Gene Expression Database (MGED) Society in producing MIAME guidelines (http://www.mged.org/Workgroups/MIAME/miame.html). More recently, this type of guidance has been modified for additional concerns related to microarray analyses for toxicology (MIAME/Tox) and for the even more complex issues involved in microarray use to examine environmental genomics (MIAME/Env). The complexity of these latter types of analyses distinguishes among field studies, conditioned field studies, and laboratory studies — all of which will require different levels of detail important to interpretation of the data.

Following the lead of the MGED Society and their MIAME standards, several groups are currently working on developing a similar set of standards that will allow for such functionality, such as the SASHIMI project (http://sashimi.sourceforge.net), the PEDRo model (Taylor et al. 2003), and the Proteomics Standards Initiative (PSI) groups (Hermjakob et al. 2004), as well as initiatives in metabolomics including the standard metabolic reporting structures (SMRS) (Lindon et al. 2005) and architecture for metabolomics (ArMet) (Jenkins et al. 2004). The goal of these activities is to establish a set of required parameters necessary for the reporting of data and deposition in public databases that will facilitate the integration of data from different sources.

1.4.6.1 Emerging Transcriptomics Resources

Table 1.3 lists some key transcriptomics resources for nonmammalian species currently used in environmental toxicology or that could serve as model surrogates for species used in environmental toxicology. It is not an exhaustive list but does illustrate the broad tools available for the nonmammalian vertebrate classes and many invertebrate phyla. Contact information is in the form of Web sites for commercial vendors or research programs or is given by e-mail for direction of inquiries.

1.4.7 PROTEOMICS

The identification of proteins within the proteome depends on a combination of different methods, each with specific advantages and limitations. Using various techniques in combination will allow researchers to elucidate the complexity of protein mixtures. It is also important to include methods amenable to analyzing post-translational modifications and identifying low-abundance proteins or proteins that have atypical characteristics, such as high compositions of basic amino acids or glycosylated or phosphorylated residues. Gel-based and non-gel-based methods have been developed and each requires high-resolution separations in 1 or more dimensions. Classical proteomics relies on separation of proteins by 2-dimensional polyacrylamide gel electrophoresis (2D-PAGE), followed by identification of proteins by mass spectrometry. This method can be performed in a high-throughput format with the help of robots to pick spots from gels, digest them with trypsin, and spot onto sample plates. Quantification of protein levels is performed in the gel by image analysis software.

The availability of the nonradioactive, fluorescent cyanine labels, Cy3 and Cy5, for proteins has enabled this technology to advance to a higher level of sophistication. Proteins from control and treated samples are each independently labeled with 1 of the dyes; the samples are combined and co-migrated in 2-dimensional gels. Each label fluoresces at a different wavelength, and each can be quantified by a specific green or red laser in a phosphorimager. This method is known as "2-dimensional differential gel electrophoresis" (2D-DIGE) and it offers a method to quantify differential expression of proteins followed by their identification by mass spectrometry. In general, 2D-DIGE works well for protein discovery and is currently the most widely used technique for detecting proteins. Although 1000 or more proteins can be visualized on a gel, it is still insufficient for complete proteome analysis. There are at least 10,000 proteins per tissue (perhaps more), and most are not visualized on the gel because of low abundance. Sample fractionation procedures are necessary to detect low-abundance proteins, making this system less high throughput than it was originally perceived to be.

The mass spectrometers of choice for proteomics analyses include matrix-assisted laser desorption ionization time-of-flight (MALDI TOF) or liquid chromatography electrospray ionization (LC-ESI) mass spectrometry. LC-ESI offers the additional advantage of obtaining sequence information from the individual tryptic fragments. Fragments can be further fractionated with liquid chromatography–mass spectrometry (LC-MS/MS) to obtain amino acid sequence information. MALDI-compatible mass spectrometers (such as quadrupole time of flight or TOF/TOF) can also generate sequence information for individual tryptic fragments. This approach is useful to identify proteins obtained from organisms whose genomes have been completely sequenced, where databases can be queried for identification by tools such as MS-Fit (http://prospector.ucsf.edu/ucsfhtml4.0/msfit.htm). For species that have little or no genome information, this application is less robust.

Potential peptide matches are possible for proteins that are highly conserved and have peptides that match homologous proteins in the database. However, this will be true only for a subset of highly conserved proteins. Without genome information, large-scale efforts to identify the majority of proteins from nonmodel systems could be accomplished with de novo protein sequencing on the mass spectrometer. The difficulty and expense of this process limits data collection; however, new instrument and technique development that will measure masses more accurately are ongoing and should help with this approach.

Non-gel-based methods have also been developed that are useful to determine differential expression of proteins in a quantitative manner. These methods rely on prelabeling proteins with isotope-coded affinity tags and mixing and digesting proteins, followed by separation of individual fragments by several dimensions of chromatography prior to analysis by mass spectrometry. These methods work well and are continually being improved for protein identification. However, this technique is limited by labeling only select proteins and is quite expensive.

The development of the cleavable isotope coded affinity tag (cICAT) refers to a pair of affinity-directed reagents that differ from each other by 9 mass units and are specific for tagging cysteine residues in proteins. After trypsin digestion, fragments from differentially expressed proteins will appear to be higher (or lower) than their corresponding partners via LC-MS/MS. Identities of the parent proteins can be obtained by comparing the sequences to databases; however, post-translational modifications may not be ascertained by this method. Newer labeling reagents are continually being developed, and some of these may be able to identify proteins that are differentially modified in response to treatment.

The development of protein "chips" employs 2 different strategies: antibody based and protein based. Antibody-based chips require antibodies to proteins of interest and would thus require genomics information or a robust database of putative coding sequences. Akin to the production of cDNA microarrays, the antibodies must be fixed to glass slides (or protein-binding membrane, nitrocellulose or polyvinylidene difluoride, PVDF) and probed with the full complement of proteins from the tissue of interest. Bound proteins may be visualized by labeling with biotin prior to hybridization, then using a secondary detection with a strepavidin-coupled alkaline phosphatase linked antibody. Alternatively, one could develop an ELISA-based detection and develop an additional set of specific antibodies for each protein. This latter method would necessitate having at least 2 antibodies for every protein, a capture antibody and a detection antibody. Hence, the availability and cost of antibodies are 2 key impediments to increasing the number of proteins using these arrays.

A similar issue for all array technologies is the standardization of protocols. Conversely, protein-based chips place proteins directly on the glass slide (or membrane) and are used primarily to test for protein–protein interactions or protein–drug interactions. The availability of purified or recombinant proteins for the species of interest would limit the application of this technique for most nonmodel species. Chip-based methods have enormous advantages in the ability to screen large portions of the proteome effectively and in the potential to be used in cross-species extrapolations where antibodies can be developed to recognize similar epitopes. However, these methods are primarily in the development phase for mammalian species and have not been applied to ecotoxicology species. Again, the main disadvantage for environmental toxicology use remains the lack of database information for nonmammalian environmental species.

There are significant limitations to the use of proteomics for comparative purposes. First, the greatest utility of proteomics is achieved when the genome is characterized. Second, the best equipment is expensive and requires highly trained operators to obtain quality information. Third, proteins can vary in concentration over several orders of magnitude in different tissues, complicating quantification and limiting global identification without prior enrichment by fractionation. For example, in the blood, protein concentrations vary over 12 orders of magnitude (e.g., cytokines such as IL-6 vs. albumin). Despite these limitations, the potential of these approaches is viewed as complementary to transcriptomics. For environmental species, a few toxicology studies to date have examined patterns of protein expression under a variety of environmental conditions (Kanaya et al. 2000; Bradley et al. 2002; Hogstrand et al. 2002; Martin et al. 2002). While patterns of differential protein expression can be assessed via 2D-PAGE and consistencies in stressor-specific profiles are evident, the identification of proteins remains a challenge.

1.4.8 METABOLOMICS — MOLECULAR PHENOTYPE AND METABOLIC TRAJECTORIES

Metabolomics directly measures the functional biochemical status of a cell or tissue within a whole organism at the molecular level; specifically, it defines the molecular phenotype that closely links to the organism physiology. This phenotype results from a complex interaction of the genotype with a multitude of environmental factors and can effectively link and integrate the changes observed by transcriptomics and proteomics. The functionality associated with metabolic profiles can be used to great benefit because it has the potential to inform of the mechanism and severity of a toxic insult. Stressors acting via specific modes of action (MOA) would be anticipated to perturb the metabolic components of a cell in a similar manner, which could then serve as a fingerprint to that specific mode of action. In principle, a library of metabolic response fingerprints could then be used for classifying the potential mode of action of uncharacterized chemicals (e.g., in toxicity tests of novel products from the chemical and pharmaceutical industries) and for identifying and characterizing the health status and exposure history of humans and wildlife following environmental exposures to toxicants.

The close linkage between metabolic data and organism physiology suggests that metabolomics will be important for risk assessment by providing molecular measurements of organism growth, survival, and fecundity. Traditionally, metabolic measurements such as adenylate energy charge and glycogen levels have been correlated with organism fitness. Although these have not proven to be particularly popular biomarkers, largely due to their lack of specificity, the measurement of multiple metabolic endpoints has the potential to assess organism health and provide mechanistic insight simultaneously.

Although the integration of genotype and environmental factors within the metabolic profile can be viewed as advantageous, it also generates a huge challenge for the field of metabolomics: to separate the metabolic fingerprints associated with specific toxicant-induced modes of action from background metabolic "noise." Indeed, after 25 years of research, the full potential of biomarkers in environmental toxicology has yet to be realized due to the variability typically encountered in individual markers. Although a component of this metabolic noise will indeed be stochastic and hence not assignable to a specific process, the majority of this variability should in principle be interpretable. For example, it is known that metabolite profiles are dictated by several phenotypic parameters (e.g., sex and age) and perturbed by a range of environmental effects including biological (e.g., disease status, parasite load, photoperiod, and seasonality of the reproductive cycle) and physical (e.g., temperature, oxygen availability, and salinity) factors.

In addition to defining the sex and age of the organism under investigation, 2 additional strategies can be used to overcome the challenges imposed by these variables. The first is to apply strict control of the environmental variables within the study, thereby minimizing the contributions from these external effectors. Such an approach has been used in laboratory-based metabolomics investigations of yeast, plants, and animals. The second strategy — the one ultimately necessary for environmental studies in which external stressors cannot be controlled — is to first characterize within the laboratory the integrated metabolic responses to toxicant exposures in the presence of single and multiple biological and physical factors. In principle, this should then enable the complex metabolic responses measured in the field to differentiate between toxicant- and nontoxicant-induced effects. Although this second approach appears daunting, many of the experimental benefits of metabolomics, including rapid and automated analyses in a cost-effective manner, reduce this to a challenging but feasible undertaking. Integral to this effort must be the continued development of classification methods for metabolomics data.

To date, the majority of metabolomics studies have examined the metabolic effects of drugs and toxicants in mammalian models (Nicholson et al. 2002). In particular, the Consortium for Metabonomic Toxicology (COMET), a collaboration among Imperial College, Bristol-Meyers Squibb, Eli Lilly and Company, Hoffman-La Roche, Novo Nordisk, and Pfizer Inc., has directed considerable funding into the development of screening tools for drug discovery using NMR technology. Previous metabolomics studies on environmentally relevant species have identified novel biomarker patterns in stressed terrestrial invertebrates, particularly earthworm species (*Eisenia veneta*), including an NMR-based investigation into the toxicity of fluorinated phenols (Bundy et al. 2001) and anilines (Bundy et al. 2002).

Recently, several applications of NMR-based metabolomics have been reported in aquatic species, including the identification of metabolic profiles associated with a muscle-withering disease in red abalone (*Haliotis rufescens*) (Viant et al. 2003) and a study of the effects of temperature stress on juvenile steelhead trout

(*Oncorhynchus mykiss*) (Viant et al. 2003). As has long been recognized by toxicologists, the molecular responses to chemical stressors change through time. By repeatedly measuring the metabolic profiles of biofluids sampled from an exposed organism, the temporal response can be reconstructed in the form of a "metabolic trajectory" that shows onset of and subsequent recovery from the insult (Nicholson et al. 2002). This concept has recently been extended to a "developmental metabolic trajectory" that has been used to characterize the NMR visible metabolome of the Japanese medaka (*Oryzias latipes*) throughout its 8-day period of embryogenesis (Viant 2003).

1.4.9 EXPERIMENTAL CONSIDERATIONS FOR METABOLOMICS

The study of metabolites is complicated by their diversity of physical and chemical properties, which vary tremendously in terms of polarities (e.g., from highly polar organic acids and bases to nonpolar lipids), concentrations (e.g., from abundant amino acids to low-concentration hormones), chemical compositions, and their redox activity and ability to ionize. Unfortunately, this range of properties precludes the use of a single analytical tool for measuring all metabolite concentrations (as can be achieved using DNA microarrays for the measurement of transcripts). Reliance on 1 toolset alone will limit (at least with current technologies) observations to only a fraction of the metabolome.

A second complication associated with the wide range of metabolite polarities arises if the metabolites need to be extracted from a tissue sample because each method will extract only a limited range of metabolite polarities. For example, acid-based extractions using perchloric acid will preferentially extract polar metabolites, whereas a methanol–chloroform extraction will isolate the polar and lipophilic metabolites in different fractions. In fact, all sample manipulations and cleanup (e.g., solid-phase extraction) have the potential to cause loss of metabolites from the sample. Therefore, careful consideration of the sample treatment and choice of bioanalytical technology is needed during the experimental design.

From the perspective of cross-species extrapolations, however, analytical methods optimized for the measurement of selected metabolites or metabolite classes can be directly applied to any species. For example, to quantify the amount of a specific fructose 1,6-biphosphatase across several species, the DNA sequence (for transcriptomics) or protein sequence with post-translational modifications (for proteomics) must be known a priori for each organism (Goodacre et al. 2004). In contrast, metabolomics methods developed to quantify fructose 1,6-bisphosphate (the reaction substrate) and fructose 6-phosphate (the product) can be used across all species because these chemicals retain the same structure across phyla.

The 2 leading platforms for metabolomics are [1H]NMR spectroscopy (Nicholson et al. 2002) and mass spectrometry (Weckwerth 2003). [1H]NMR-based methods have multiple benefits that include

- rapid and simple sample preparation that minimizes the loss of metabolites compared with more rigorous sample clean-up protocols,
- global observation of all high-abundance metabolites that contain nonexchangeable hydrogen atoms in a single "all-in-one" analysis,

- quantitative metabolite measurements with a high degree of reproducibility,
- automated analyses (>100 samples/day), and
- robust and established NMR technology with minimal instrument down-time inexpensive on a per-sample basis.

This last point is critical because, on a consumables basis, literally many tens of samples can be analyzed for the price of 1 transcriptomics replicate, which enables large metabolomics studies to be conducted in a rapid and cost-effective manner. The primary disadvantage of this approach is the relatively poor sensitivity of NMR spectroscopy, which has limited capability to detect concentrations in the nanogram per milliliter level. This limits observation under relatively standard conditions to typically between 100 and 200 metabolites from several different classes, including amino acids, organic acids and bases, nucleotides, and carbohydrates. This amounts to less than 10% of a typical metabolome. In summary, NMR-based metabolomics provides an unbiased top-down (or screening) approach that can provide insight into a broad array of metabolism. The limitation of this technology is the complexity of the spectrum of which only a small fraction of the peaks can be associated with a known metabolite.

Mass spectrometry is a significantly more sensitive analytical tool and can directly complement the use of NMR spectroscopy in metabolomics by facilitating the analysis of less abundant metabolites, as depicted in Figure 1.2. Although mass spectrometry shares several of the advantages of the NMR approach (e.g., capable of automation, relatively inexpensive on a consumables basis), several additional technical issues must be considered:

1) Metabolites will only be detected if they can be ionized, thus limiting the global profiling capabilities.
2) While direct injection of the sample into the mass spectrometer is possible, typically, the instrument is coupled to a GC or LC, which increases the analysis time but benefits from the resolution of metabolic profiles into a second dimension.
3) Quantification of multiple metabolites is currently difficult (although this could be solved by inclusion of metabolite standards enriched with stable isotopes).
4) The experimental configuration is somewhat more susceptible to instrument downtime compared with NMR.

These challenges have limited the introduction of mass spectrometry into metabolomics-based toxicological studies compared with NMR methods, although a few good studies have recently been reported (Lenz et al. 2004; Williams et al. 2004). Mass spectrometry-based metabolomics can therefore provide a bottom-up (or targeted) approach that can selectively measure (all) the metabolites associated with a specific metabolic pathway. Overall, the combination of the breadth of the metabolic profile obtained by NMR with the sensitivity and chemical characterization provided by mass spectrometry offers a powerful approach to identify biomarker profiles of

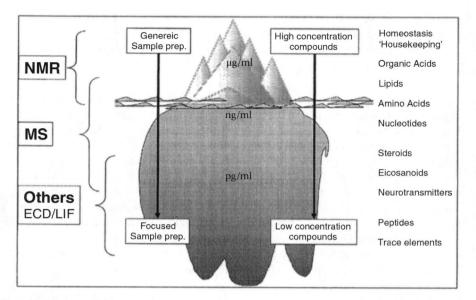

FIGURE 1.2 "Iceberg" cartoon illustrating that comprehensive coverage of the metabolome requires an array of analytical techniques and different sample preparation strategies to encompass the diverse range of physical and chemical properties of cellular metabolites. Listed are various techniques used in metabolomics (left column) and representative metabolite classes (right column). All metabolites span a range of concentrations, and the delineations in the central column are meant to be qualitative in nature. ECD/LIF is an acronym of electrchemical detection/laser-induced fluorescense. (From Harrigan, G. and Goodacre, R., Eds. 2003. *Metabolic Profiling — Its Role in Biomarker Discovery and Gene Function Analysis.* Dordrecht: Kluwer Academic Publishers, 174. With permission.)

toxic insult. Unfortunately, there are few examples to illustrate this promise due to the newness of this technology.

1.4.10 ANNOTATION OF CELLULAR METABOLOME

One of the biggest problems for all the analytical methods used in metabolomics is poor annotation of the observed peaks. In a typical NMR study, fewer than 30 metabolites of the estimated few hundred within the spectrum will be positively identified. This limits the metabolic information that can be extracted from the spectral data and restricts the mechanistic insight that could potentially be gained. Furthermore, annotation is a prerequisite for instrument-independent metabolic data, which are the desired format for the construction of metabolic databases and facilitate the comparison of datasets regardless of the instrument on which they were measured. If metabolomics is to fulfill its potential, this problem must be addressed urgently by construction of metabolite libraries that contain ^1H and ^{13}C NMR spectra of hundreds of metabolite standards and mass spectra of metabolites that have been labeled with stable isotopes such as ^{13}C (to aid annotation and quantification of unknown peaks). Ultimately, metabolite libraries will be of great benefit for cross-species toxicological studies, as well as to all medical, nutritional, biological, and environmental metabolomics. Unfortunately, no consortia or groups within the public sector are pursuing the creation of metabolite libraries at present.

1.5 PATHWAY MAPPING — THE FUTURE OF OMICS TECHNOLOGIES

Pathway mapping and the bioinformatic approaches are described in detail in Chapter 4 in this book. Historically, single biomarkers have been used for disease and toxicity assessment. For example, elevated glucose has long been used as an indicator of diabetes. However, this approach will always fail to identify why a particular marker is altered. Furthermore, not only are single biomarkers limited in their ability to inform on mechanism, but variability within a single marker can also reduce the likelihood of classifying an organism in a particular state. An ideal approach would be to identify a profile of biomarkers that contains sufficient information to 1) positively identify the occurrence of toxicity and pathology against a background of biological and environmental variability and 2) identify the mode of action by which the toxicant is acting (which in turn could suggest the class of toxicant if this information is unknown, as is often the case in ecotoxicology). This approach would be particularly powerful if all the components associated with a specific mode of toxicity or pathway (e.g., oxidative stress or endocrine disruption) — termed "pathway mapping" — could be compared across multiple species; in principle, this approach is already achievable using metabolomics.

Consequently, it would be a mistake to apply omics technology only as a discovery-driven approach in order to identify a (novel) single biomarker associated with a particular disease or toxicity endpoint. The power of these approaches stems from the information contained within the profile. An even more robust profile would integrate, for example, transcriptional and metabolic markers, thus combining the high-specificity and regulatory information associated with transcriptomics measurements with the high functionality of metabolic data. It is important to note that, contrary to popular belief, these omics approaches do not provide greater sensitivity compared with other similar molecular techniques. In fact, the measurement of multiple endpoints is often accompanied by an overall loss of sensitivity compared to a method that is optimized for 1 particular endpoint. This is a small price to pay, however, for the vastly increased amount of information that can be mined from omics data. For instance, the comprehensive characterization provided by omic approaches complements and enhances the ability to detect patterns of gene alterations that precede toxicity and facilitate identification of biomarkers.

1.6 EXAMPLE OF CROSS-SPECIES EXTRAPOLATION USING TRANSCRIPTOMICS

As omics technologies develop for various species, the opportunity for efficiently retrieving information from 1 species and using it to better understand another species has arrived. Informatic approaches that can facilitate cross-species comparisons are described in Chapter 4 and elsewhere in this book. Gene ontology designations

(Consortium GO 2000) allow a comparison of similar functions of genes from different species. Programs that allow the construction of pathways from microarray data are developed and being developed (Dahlquist et al. 2002) and should allow cross-species extrapolations of microarray data into organized, functional pathways. Attempts at comparing families of genes in different species have begun (Stuart et al. 2003) and, as respective technologies develop, this comparison will occur with greater frequency.

Biologically, using traditional and nontraditional techniques, one finds growing numbers of comparable signaling pathways in different species that justify careful omics comparisons. While these comparisons have yet to be done in a comprehensive manner, there are several strong candidates for such investigation. One, retinoid signaling, is of critical importance to proper embryonic development of vertebrates (Maden 1999). Vitamin A (retinol) is obtained from our diets, and decades of research have shown that vitamin A excess or deficiency in many different vertebrate embryos can cause several severe organogenesis defects. The operant molecule in this process is the acid form of vitamin A, retinoic acid (RA). RA functions as a ligand for retinoic acid receptors (RARs), which are ligand-activated transcription factors in the steroid receptor superfamily.

Recently, several studies in mouse have shown that the limiting factors in proper retinoid signaling in embryos are the retinaldehyde dehydrogenase 2 gene (Raldh2), which produces retinoic acid from retinaldehyde (retinaldehyde is produced by the action of retinol dehydrogenase on vitamin A), and a series of genes in the cytochrome P450 family (CYP26a1, b1, and c1). These CYP26 enzymes metabolize retinoic acid to a nonfunctional form (Niederreither, Abu-Abed, et al. 2002). The Raldh2 and CYP26a1 enzymes are expressed in localized regions of the embryo, usually in close proximity to one another, and their opposing activities localize RA in functionally significant regions and times of the developmental process (Niederreither et al. 1997; Swindell et al. 1999; Dobbs-McAuliffe et al. 2004).

In comparing vertebrate species such as mouse with zebrafish, one finds remarkable parallels in retinoid signaling, the expression of Raldh2 and the CYP26s, and extreme overlap in comparable mutants between the 2 species. Transgenic indicator mice and zebrafish for retinoic acid activity illustrate gene expression in common regions and developmental sequence (Rossant et al. 1991; Balkan et al. 1992; Perz-Edwards et al. 2001) (Figure 1.3). Mouse and zebrafish mutants in Raldh2 show common phenotypes, including hindbrain and forelimb defects (Niederreither et al. 1999; Begemann et al. 2001; Grandel et al. 2002; Niederreither et al. 2002). CYP26a1 knockout mutants in mice create phenotypes including spina bifida to caudal truncations (Abu-Abed et al. 2001; Sakai et al. 2001). These are consistent with the expression patterns of CYP26a1 in the growing caudal end of mouse and zebrafish embryos (Abu-Abed et al. 2002; Dobbs-McAuliffe et al. 2004).

The phenotypes of the mouse mutants suggest that the control of retinoid levels during development is critical to proper development. Therefore, environmental influences that might affect the expression or activity of Raldh or CYP26 could create developmental abnormalities through the repositioning of RA. Recently, in model systems as diverse as human epithelial airway cells in culture, there have been reports that the 2,3,7,8-tetrachlorodibenzo-*D*-dioxin (TCDD) form of dioxin

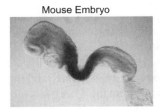

Mouse Embryo   Zebrafish Embryo

Cross-species RA Elements:

FIGURE 1.3 Cross-species comparisons between mouse (left panel) and zebrafish (right panel) embryos for the retinoic acid (RA) receptor. Both species have RA receptors, retinoid-X receptors, Raldh2s, and CYP26 isoforms. Mutants in Raldh2 have similar phenotypic malformations. This figure illustrates the concordance across species in receptor and enzyme targets. This concordance is a premise for using omics technologies in facilitating cross-species extrapolations.

(which can act through the aryl hydrocarbon receptor system) can affect the expression of genes in the retinoid pathway (Gambone et al. 2002; Martinez et al. 2002). Therefore, the possibility exists that fetal exposure to environmental toxicants could impose developmental changes through this retinoid signaling pathway.

In conceptualizing and experimentally attacking these issues through genomics technologies and, in particular, microarray analysis, the parallels between zebrafish and mouse allow one to use some of the advantages of this complementary vertebrate system for higher throughput analysis of toxicant interactions. For example, microarray analysis of toxicant exposure coupled with selective gene knockdown through gene-specific morpholino introduction into embryos can provide connections and mechanistic understanding that can then be more efficiently applied to mammalian models. Such cross-species comparisons using microarray analysis in combination with other approaches should allow more efficient dissection of environmental and toxicant effects upon vertebrate development.

One of the advantages of considering different model systems in toxicology is the ability to pick and choose systems to overcome technical hurdles of 1 system and exploit another. One advantage of fish systems, which lay their eggs for fertilization in the water, is the opportunity to perform experiments on developing embryos outside the mother, thus allowing the visualization of events as they occur. Technology around zebrafish has developed to a high degree and, besides visualization, a growing number of fluorescent, transgenic lines have been developed. These allow one to specifically isolate and visualize parts of the nervous system, the vasculature, and the nuclei of all cells, as well as to examine the response of embryos to small molecules through transgenics whose transcription can be controlled by small molecules added to the water.

When this technology is coupled with microarray analysis, it is possible to examine continuous changes of specific regions of embryos via fluorescence or confocal analysis in 4 dimensions. Systems such as these allow the introduction of gene-specific antisense morpholinos that allow the knockdown translation of genes for approximately 3 days, the time period of embryogenesis in zebrafish. From a neurotoxicological standpoint (Linney et al. 2004), the neural tube of zebrafish develops within 24 hours, response to touch occurs within 21 hours, and a swim response from touch develops within 27 hours (Saint-Amant and Drapeau 1998). Therefore, the response of the embryo and resulting larvae to environmental toxicants can be surveyed visually throughout these developmental time periods, and the effects of inhibiting the translation of specific genes can be examined and monitored with fluorescent transgenics. Sophisticated software tracking systems can now collect video images of movement and motion over time and be queried to extract quantifiable differences from swimming embryos and larvae for graphing and comparison (Peitsaro et al. 2003).

If one couples microarray analysis with environmental exposure, fluorescent transgenic analysis (Udvadia and Linney 2003), and tracking data for the analysis of potential changes imposed upon the organisms by earlier exposure, a synergistic approach can be developed that would aid in identifying critical developmental windows of vulnerability with the resulting phenotypic change on larval behavior. A prototypical version of this approach is being developed where chlorpyrifos exposure of zebrafish from 0 to 5 days postfertilization results in adults with learning deficiencies (Levin et al. 2003). A preliminary report identifies larval motility differences after similar exposures (Levin et al. 2004). This work analyzed motility manually, but these changes have now been confirmed using the Noldus–Ethovision tracking system (Peitsaro et al. 2003). This type of analysis will reduce the time needed to characterize the window of vulnerability for the behavioral changes produced by the pesticide.

This combination of approaches should allow an identification of a narrow window of developmental exposure that produces a behavioral phenotype. These approaches can be coupled with microarray analysis for the examination of gene changes occurring during exposure. Cluster analysis of gene changes over time can then potentially identify pathways that are critically important to the resultant phenotype. It is possible to use a variety of technologies in an integrated and interactive way to study the whole organism by combining fluorescent transgenics to examine gene expression changes or structural changes in the embryo in response to exposure, microarray analyses to survey specific gene expression changes occurring during exposure, behavioral changes with tracking software shortly after exposure, and long-term learning ability in adults. These technologies can then be combined to focus upon critical changes induced by the environmental and toxicant exposure with the possibility of developing information that would then allow a more efficient and complementary examination in rodent models.

The value of taking advantage of different model systems is that they allow a higher throughput analysis of environmental exposure. Recent studies of chemicals in meconium from human fetuses in the Philippines (Ostrea et al. 2002) have identified the presence of lead, cadmium, mercury, chlordane, lindane, and chlorpyrifos. Therefore, the distinct possibility exists that human fetuses are being exposed to chemicals that might affect learning and behavior of children and adults. Having alternative approaches and models such as these could help assess whether there are significant changes in the gene expression in these exposed populations.

1.7 RECOMMENDATIONS

Our workgroup identified several obstacles that impede the use of omics technologies in cross-species extrapolations. The following recommendations would overcome these impediments:

- Facilitate cross-species extrapolations using omics technologies by coordinating an international research agenda funded through government agencies in order to ensure the continued development of technologies and standardized protocols and to address critical data gaps. It is recommended that a joint SETAC-SOT task force be created to outline an international research agenda that could be used by funding agencies in their priority setting.
- Establish a public-domain international library of NMR spectra of metabolite and mass spectra of stable isotope-labeled metabolites, using standardized protocols, to aid in the annotation and quantification of metabolomics data.
- Establish a public-domain international library of genomics polymorphisms and transcriptomics, proteomics, and metabolomics response fingerprints employing model compounds in key ecotoxicology models used for safety assessment that are phenotypically anchored to whole-organism and population-level changes.
- Support a panomic case study in pathway mapping across several species used in safety assessment.
- Support hypothesis- and discovery-driven research in environmental toxicology using omics technologies.
- Focus research on genome sequencing and annotation of key environmental toxicology models used in safety assessment that will facilitate cross-species extrapolations.
- Support cross-laboratory validation efforts to identify and improve protocols and reduce interlaboratory variability as well as understand the normal range of variability in expression profiles.

1.8 FUTURE

The future state is envisioned to be the following within 5 to 10 years:

- DNA microarrays of key ecotoxicology species are commercially available and comprise transcripts of known identity and function.
- Peaks within proteomics and metabolomics fingerprints can be assigned to known proteins and metabolites, respectively. Improved technologies will provide greater coverage of the cellular proteome and metabolome.

- Integration of genomics, transcriptomics, proteomics, and metabolomics measurements can be achieved, thus providing comprehensive molecular responses to toxicant exposure. These molecular events will be phenotypically anchored to whole-organism and population-level effects.
- A database comprising landscapes of RNA, protein, and metabolite markers associated with established MOAs in model organisms would be used to determine potentially adverse effects of new products from the chemical or pharmaceutical industries. This would allow identification of perturbed transcriptional, protein, or metabolic pathways and networks and provide visualization of the MOAs based upon, for example, the KEGG database. Ideally, these databases would include information across a range of dose levels to define dose–response patterns.
- Ultimately, perhaps within 10 years, genomics polymorphisms and transcriptomics, proteomics, and metabolomics fingerprints obtained from environmentally sampled organisms could be compared to database libraries to identify potential exposure profiles indicative of impending adverse organism outcomes and population declines.

REFERENCES

Abu-Abed S, Dolle P, Metzger D, Beckett B, Chambon P, Petkovich M. 2001. The retinoic acid-metabolizing enzyme, CYP26A1, is essential for normal hindbrain patterning, vertebral identity, and development of posterior structures. Genes Dev 15:226-240.

Abu-Abed S, MacLean G, Fraulob V, Chambon P, Petkovich M, Dolle P. 2002. Differential expression of the retinoic acid-metabolizing enzymes CYP26A1 and CYP26B1 during murine organogenesis. Mech Dev 110:173–177.

Alizadeh AA, Eisen MB, Davis RE, Ma C, Lossos IS, Rosenwald A, Boldrick JC, Sabet H, Tran T, Yu X, and others. 2000 Distinct types of diffuse large B-cell lymphoma identified by gene expression profiling. Nature 403:503–511.

Alon U, Barkai N, Notterman DA, Gish K, Ybarra S, Mack D, Levine AJ. 1999. Broad patterns of gene expression revealed by clustering analysis of tumor and normal colon tissues probed by oligonucleotide arrays. Proc Natl Acad Sci U S A 96:6745–6750.

Balkan W, Colbert M, Bock C, Linney E. 1992. Transgenic indicator mice for studying activated retinoic acid receptors during development. Proc Natl Acad Sci U S A 89(8):3347–3351.

Begemann G, Schilling TF, Rauch GJ, Geisler R, Ingham PW. 2001 The zebrafish neckless mutation reveals a requirement for raldh2 in mesodermal signals that pattern the hindbrain. Development 128(16):3081–3094.

Bradley B., Shrader EA, Kimmel DG, Meiller JC. 2002. Protein expression signatures: an application of proteomics. Mar Environ Res 54:373–377.

Brazma A, Hingamp P, Quackenbush J, Sherlock G, Spellman P, Stoeckert C, Aach J, Ansorge W, Ball CA, Causton HC, and others. 2001. Minimum information about a microarray experiment (MIAME)-toward standards for microarray data. Nat Genet 29:365–371.

Bundy JG, Osborn D, Weeks JM, Lindon JC, Nicholson JK. 2001. An NMR-based metabonomic approach to the investigation of coelomic fluid biochemistry in earthworms under toxic stress. FEBS Lett 500:31–35.

Bundy JG, Lenz EM, Bailey NJ, Gavaghan CL, Svendsen C, Spurgeon D, Hankard PK, Osborn D, Weeks JM, Trauger SA. 2002. Metabonomic assessment of toxicity of 4-fluoroaniline, 3,5-difluoroaniline and 2-fluoro-4-methylaniline to the earthworm *Eisenia veneta* (Rosa): Identification of new endogenous biomarkers. Environ Toxicol Chem 21:1966–1972.

Burczynski ME, McMillian M, Ciervo J, Li L, Parker JB, Dunn RT 2nd, Hicken S, Farr S, Johnson MD. 2000. Toxicogenomics-based discrimination of toxic mechanism in HepG2 human hepatoma cells. Toxicol Sci 58:399–415.

Consortium GO. 2000. Gene ontology: tool for the unification of biology. Nat Genet 25:25–29.

Dahlquist KD, Salomonis N, Vranizan K, Lawlor SC, Conklin BR. 2002. GenMAPP, a new tool for viewing and analyzing microarray data on biological pathways. Nat Genet 31:19–20.

Daly AK. 2003. Pharacogenetics of the major polymorphic metabolizing enzymes. Fundamental Clin Pharmacol 17:27–41.

Dobbs-McAuliffe B, Zhao Q, Linney E. 2004. Feedback mechanisms regulate retinoic acid production and degradation in the zebrafish embryo. Mech Development 121:339–350.

Eaton DL, Klaassen CD. Principles of toxicology. In: Klaassen CD, editor. Casarett and Doul's toxicology: the basic science of poisons. 6th ed. New York: McGraw Hill; 2001, p 11–34.

Enard W, Khaitovich P, Klose J, Zollner S, Heissig F, Giavalisco P, Nieselt-Struwe K, Muchmore E, Varki A, Ravid R, et al. 2002. Intra- and interspecific variation in primate gene expression patterns. Science 296(5566):340–343.

Farrall M. 2004. Quantitative genetic variation: a post-modern view. Hum Mol Genet 13(1):R1–R7.

Gambone CJ, and others. 2002. Unique property of some synthetic retinoids: activation of the aryl hydrocarbon receptor pathway. Mol Pharmacol 61:334–342.

Goodacre R, Vaidyanathan S, Dunn WB, Harrigan GG, Kel, DB. 2004. Metabolomics by numbers: acquiring and understanding global metabolite data. Trends Biotechnol 22: 245–252.

Grandel H, Lun K, Rauch GJ, Rhinn M, Piotrowski T, Houart C, and others. 2002. Retinoic acid signalling in the zebrafish embryo is necessary during pre-segmentation stages to pattern the anterior-posterior axis of the CNS and to induce a pectoral fin bud. Development 129(12):2851–2865.

Hacia JG. 1999. Resequencing and mutational analysis using oligonucleotide microarrays. Nat Genet 21(1 Suppl):42–47.

Hamadeh HK, Bushel PR, Jayadev S, DiSorbo O, Bennett L, Li L, Tennant R, Stoll R, Barrett JC, Paules RS, and others. 2002. Prediction of compound signature using high density gene expression profiling. Toxicol Sci 67:232–240.

Hamadeh HK, Bushel PR, Jayadev S, Martin K, DiSorbo O, Sieber S, Bennett L, Tennant R, Stoll R, Barrett JC, and others. 2002. Gene expression analysis reveals chemical-specific profiles. Toxicol Sci 67:219–231.

Hermjakob H, Montecchi-Palazzi L, Bader G, Wojcik J, Salwinski L, Ceol A, Moore S, Orchard S, Sarkans U, von Mering C, et al. 2004. The HUPO PSI's molecular interaction format—a community standard for the representation of protein interaction data. Nat Biotechnol 22:177–183.

Hoekenga OA, Vision TJ, Shaff JE, Monforte AJ, Lee GP, Howell SH, Kochian LV. 2003. Identification and characterization of aluminum tolerance loci in *Arabidopsis* (*Landsberg erecta* x *Columbia*) by quantitative trait locus mapping. A physiologically simple but genetically complex trait. Plant Physiol 132(2):936–948.

Hogstrand C, Balesaria S, Glover CN. 2002. Application of genomics and proteomics for study of the integrated response to zinc exposure in a non-model fish species, the rainbow trout. Comp Biochem Physiol B Biochem Mol Biol 133:523–535.

Hughes TR, Marton MJ, Jones AR, Roberts CJ, Stoughton R, Armour CD, Bennett HA, Coffey E, Dai H, He YD, and others. 2000. Functional discovery via a compendium of expression profiles. Cell 102:109–126.

Innocenti F, Ratain MJ. 2002. Update on pharmocogentics in cancer chemotherapy. Eur J Cancer 38:639–644.

Jansen RC, Nap JP. 2001. Genetical genomics: the added value from segregation. Trends Genet 17(7):388–391.

Jenkins H, Hardy N, Beckmann M, Draper J, Smith AR, Taylor J, Fiehn O, Goodacre R, Bino RJ, Hall R, Kopka J, and others. 2004. ArMet; a proposed framework for the description of plant metabolomics experiments and their results. Nat Biotechnol 22:1601–1606.

Kanaya S, Ujiie Y, Hasegawa K, Sato T, Imada H, Kinouchi M, Kudo Y, Ogata T, Ohya H, Kamada H, and others. 2000. Proteome analysis of Oncorhynchus species during embryogenesis. Electrophoresis 21:1907–1913.

King MC, Wilson AC. 1975. Evolution at two levels in humans and chimpanzees. Science 188(4184):107–116.

Kirk BW, Feinsod M, Favis R, Kliman RM, Barany F. 2002. Single nucleotide polymorphism seeking long term association with complex disease. Nucleic Acids Res 30(15): 3295–3311.

Knoebl I, Hemmer MJ, Denslow ND. 2004. Induction of zona radiata and vitellogenin genes in estradiol and nonylphenol exposed male sheepshead minnows (Cyprinodon variegatus). Mar Environ Res 58(2-5):547–551.

Koza-Taylor PH, Cook JC, Cappon GD, Deng S, Obourn JD, Lawton MP. 2005. Global expression profiling of male rat kidney: Co-administration of a specific estrogen receptor (ER) antagonist inhibits a dynamic 17β-estradiol (E2) response. Toxicologist 90(S1):1802.

Larkin P, Folmar LC, Hemmer MJ, Poston AJ, Denslow ND. 2003. Expression profiling of estrogenic compounds using a sheepshead minnow cDNA macroarray. EHP Toxicogenom 111(1T):29–36.

Lenz EM, Bright J, Knight R, Wilson ID, Major H. 2004. A metabonomic investigation of the biochemical effects of mercuric chloride in the rat using H-1 NMR and HPLC-TOF/MS: time dependant changes in the urinary profile of endogenous metabolites as a result of nephrotoxicity. Analyst 129:535–541.

Levin E, Chrysthansis E, Yacisin K, Linney E. 2003. Chlorpyrifos exposure of developing zebrafish: effects on survival and long-term effects on response latency and spatial discrimination. Neurotoxicol Teratol 25:51–57.

Levin ED, Swain, HA, Donerly, S. Linney, E. 2004. Developmental chlorpyrifos effects on hatchling zebrafish swimming behavior. Neurotoxicol Teratol 26:719–723.

Lindon JC, Nicholson JK, Holmes E, Keun HC, Craig A, Pearce JTM, Bruce SJ, Hardy N, Sansone SA, and others. 2005. SMRS; summary recommendations for standardization and reporting of metabolic analyses. Nature Biotechnol 23(7):833–838.

Linney E, Upchurch L, Donerly S. 2004. Zebrafish as a neurotoxicological model. Neurotoxicol Teratol 26:709–718.

Luikart G, England PR, Tallmon D, Jordan S, Taberlet P. 2003. The power and promise of population genomics: from genotyping to genome typing. Nat Rev Genet 4(12):981–994.

Lynch M, Walsh B. 1997. Genetics and analysis of quantitative traits. Sunderland, MA: Sinauer Associates.

Maden M. 1999 Heads or tails? Retinoic acid will decide. Bioessays 21:809–812.

Martin SAM, Blaney S, Bowman AS, Houlihan DF. 2002. Uniquitin-proteasome-dependent proteolysis in rainbow trout (*Oncorhynchus mykiss*): effect of food deprivation. Eur J Physiol 445:257–266.

Martin R. Leppert D. 2004. A plea for "Omics" research in complex diseases such as multiple sclerosis—a change of mind is needed. J Neurol Sci 222:3–5.

Martinez JM, and others. 2002. Differential toxicogenomic responses to 2,3,7,8-tetrachlorod-ibenzo-p-dioxin in malignant and nonmalignant human airway epithelial cells. Toxicol Sci 69:409–423.

Marton MJ, DeRisi JL, Bennett HA, Iyer VR, Meyer MR, Roberts CJ, Stoughton R, Burchard J, Slade D, Dai H, and others. 1998. Drug target validation and identification of secondary drug target effects using DNA microarrays [see comments]. Nat Med 4:1293–1301.

Nicholson JK, Connelly J, Lindon JC, Holmes E. 2002. Metabonomics: a platform for studying drug toxicity and gene function. Nat Rev Drug Discov 1:153–161.

Niederreither K, McCaffery P, Drager UC, Chambon P, Dolle P. 1997. Restricted expression and retinoic acid-induced downregulation of the retinaldehyde dehydrogenase type 2 (RALDH-2) gene during mouse development. Mech Dev 62:67–78.

Niederreither K, Subbarayan V, Dolle P, Chambon P. 1999. Embryonic retinoic acid synthesis is essential for early mouse post- implantation development. Nat Genet 21(4):444–448.

Niederreither K, Abu-Abed S, Schuhbaur B, Petkovich M, Chambon P, Dolle P. 2002a. Genetic evidence that oxidative derivatives of retinoic acid are not involved in retinoid signaling during mouse development. Nat Genet 31:84–88.

Niederreither K, Vermot J, Schuhbaur B, Chambon P, Dolle P. 2002b. Embryonic retinoic acid synthesis is required for forelimb growth and anteroposterior patterning in the mouse. Development 129(15):3563–3574.

[OECD] Organization for Economic Cooperation and Development. 2004. OECD report of initial work towards the validation of a fish screening assay for the detection of endocrine-active substances: Phase 1A. Paris, France.

Oleksiak MF, Churchill GA, Crawford DL. 2002. Variation in gene expression within and among natural populations. Nat Genet 32(2):261–266.

Ostrea EM Jr, Morales V, Ngoumgna E, Prescilla R, Tan E, Hernandez E, and others. 2002. Prevalence of fetal exposure to environmental toxins, as determined by meconium analysis. Neurotoxicology 23:329–339.

Peakall D, Shugart L. 2002 The Human Genome Project (HGP). Ecotoxicology 11(1):7.

Peitsaro N, Kaslin J, Anickhtchik OV, Panula P. 2003. Modulation of the histamineric system and behavior by alpha-fluoromethylhistidine in zebrafish. J Neurochem 86:432–441.

Perou CM, Jeffrey SS, van de Rijn M, Rees CA, Eisen MB, Ross DT, Pergamenschikov A, Williams CF, Zhu SX, Lee JC, et al. 1999. Distinctive gene expression patterns in human mammary epithelial cells and breast cancers. Proc Natl Acad Sci USA 96:9212–9217.

Perz-Edwards A, Hardison NL, Linney E. 2001. Retinoic acid-mediated gene expression in transgenic reporter zebrafish. Dev Biol 229(1):89–101.

Rockett JC, Hellmann GM. 2004. Confirming microarray data—is it really necessary? Genomics 83:541–549.

Riordan JR, Rommens JM, Kerem B, Alon N, Rozmahel R, Grzelczak Z, Zielenski J, Lok S, Plavsic N, Chou JL. 1989. Identification of the cystic fibrosis gene: cloning and characterization of complementary DNA. Science 245(4922):1066–1073.

Rossant J, Zirngibl R, Cado D, Shago M, Giguere V. 1991. Expression of a retinoic acid response element-hsplacZ transgene defines specific domains of transcriptional activity during mouse embryogenesis. Genes Develop 5:1333–1344.

Saint-Amant L, Drapeau P. 1998. Time course of the development of motor behaviors in the zebrafish embryo. J Neurobiol 37(4):622–632.

Sakai, Y, Meno C, Fujii H, Nishino J, Shiratori H, Saijoh Y, Rossant J, Hamada H. 2001. The retinoic acid-inactivating enzyme CYP26 is essential for establishing an uneven distribution of retinoic acid along the anterio-posterior axis within the mouse embryo. Genes Dev 15:213–225.

Sen S, Ferdig M. 2004. QTL analysis for discovery of genes involved in drug responses. Curr Drug Targets Infect Disord 4(1):53–63.

Sone K, Hinago M, Kitayama A, Morokuma J, Ueno N, Watanabe H, Iguchi T. 2004. Effects of 17beta-estradiol, nonylphenol, and bisphenol-A on developing Xenopus laevis embryos. Gen Comp Endocrinol 138(3):228–236.

Stuart JM, Segal E, Koller D, Kim SK. 2003. A gene-coexpression network for global discovery of conserved genetic modules. Science 302:249–55.

Swindell EC, Thaller C, Sockanathan S, Petkovich M, Jessell TM, Eichele G. 1999. Complementary domains of retinoic acid production and degradation in the early chick embryo. Dev Biol 216:282–296.

Taylor CF, Paton NW, Garwood KL, Kirby PD, Stead DA, Yin Z, Deutsch EW, Selway L, Walker J, Riba-Garcia I, and others. 2003. A systematic approach to modeling, capturing, and disseminating proteomics experimental data. Nat Biotechnol 21:247–254.

Thomas RS, Rank DR, Penn SG, Zastrow GM, Hayes KR, Pande K, Glover E, Silander T, Craven MW, Reddy JK, and others. 2001. Identification of toxicologically predictive gene sets using cDNA microarrays. Mol Pharmacol 60:1189–1194.

Thomas RS, Rank DR, Penn SG, Zastrow GM, Hayes KR, Hu T, Pande K, Lewis M, Jovanovich SB, Bradfield CA. 2002. Application of genomics to toxicology research. Environ Health Perspect 110(Suppl 6):919–923.

Udvadia A, Linney E. 2003. Windows into development: historic, current, and future perspectives on transgenic zebrafish. Develop Biol 256:1–17.

[USEPA] US Environmental Protection Agency. 1998. Endocrine disruptor screening and testing advisory committee (EDSTAC) report. Washington, DC. Office of Prevention, Pesticides and Toxic Substances.

Viant MR, Rosenblum ES, Tjeerdema RS. 2003a. NMR-based metabolomics: a powerful approach for characterizing the effects of environmental stressors on organism health. Environ Sci Technol 37:4982–4989.

Viant MR, Werner I, Rosenblum ES, Gantner AS, Tjeerdema RS, Johnson ML. 2003b. Correlation between heat-shock protein induction and reduced metabolic condition in juvenile steelhead trout (Oncorhynchus mykiss) chronically exposed to elevated temperature. Fish Physiol Biochem 29:159–171.

Viant MR. 2003c. Improved methods for the acquisition and interpretation of nmr metabolomic data. Biochem Biophys Res Comm 310:943–948.

Vitt U, Gietzen D, Stevens K, Wingrove J, Becha S, Bulloch S, Burrill J, Chawla N, Chien J, Crawford M, et al. 2004. Identification of candidate disease genes by EST alignments, synteny, and expression and verification of Ensembl genes on rat chromosome 1q43-54. Genome Res 14(4):640–650.

Waring, Ciurlionis, et al. 2001. Microarray analysis of hepatotoxins in vitro reveals a correlation between gene expression profiles and mechanisms of toxicity. Toxicol Lett 120:359–368.

Waring, Jolly, et al. 2001. Clustering of hepatotoxins based on mechanism of toxicity using gene expression profiles. Toxicol Appl Pharmacol 175:28–42.

Waters M, Boorman G, Bushel P, Cunningham M, Irwin R, Merrick A, Olden K, Paules R, Selkirk J, Stasiewicz S, et al. 2003. Systems toxicology and the Chemical Effects in Biological Systems (CEBS) knowledge base. Environ Health Perspect 111:811–824.

Waters MD, Fostel JM. 2004. Toxicogenomics and system toxicology: aims and prospects. Nat Rev Genet 5(12):936–948.

Weckwerth W. 2003. Metabolomics in systems biology. Annu Rev Plant Biolo 54:669–689.

Williams RE, Lenz EM, Major HJ, Lock EA, Wilson ID. 2004. D-serine-induced nephrotoxicity: a HPLC-TOF/MS metabonomics approach. Toxicology 202:65–66.

Williams TD, Gensberg K, Minchin SD, Chipman JK. 2003. A DNA expression array to detect toxic stress response in European flounder (*Platichthys flesus*). Aquat Toxicol 65(2):141–57.

2 Selection of Surrogate Animal Species for Comparative Toxicogenomics

Nancy D. Denslow, John K. Colbourne, David Dix, Jonathan H. Freedman, Caren C. Helbing, Sean Kennedy, and Phillip L. Williams

2.1 INTRODUCTION

The goal of this chapter is to identify and prioritize species of animals that can serve as surrogates for environmental and human health in comparative toxicogenomics studies. By characterizing current research needs and by deriving a set of criteria for selecting appropriate surrogate species, we will accomplish this goal. These criteria will be applied in reviewing the species currently used in toxicological testing and research, ecological monitoring and research, and the genomic sciences. The review includes a short discussion on why a community-based approach is used (in one case) to promote a model species for this emerging field and will conclude by identifying a suite of species that could serve as an initial starting point for a multinational comparative toxicogenomics research program.

For each taxonomic class, the utility of the putative surrogate species as sentinels for environmental and human health must be considered. This condition will optimize their comparative value to obtain complementary results. Especially useful are species that cross boundaries between various environments (e.g., aquatic or terrestrial) and toxicities and that represent a model for diseases and outcomes outside their class. Genomics greatly enables this comparative approach by using the genetic code as an evolutionarily conserved common link between species and by connecting the ecological and health domains. Genomics information will allow quantification of known commonalities in response to and in discovery of previously unrecognized toxicity pathways shared between species. Furthermore, genomics resources will fuel highly parallel assays (e.g., DNA microarrays) and inform the interpretation of proteomics and metabolomics datasets, which will greatly enhance the pace of scientific progress.

The identification of additional toxicological test species is necessary to bridge the gaps between current test species, which are critical for understanding the effects of chemical and environmental stressors, and species for which genomes have been sequenced. The current test and monitoring species used around the globe have generated an enormous amount of data and have been developed for various practical and historical reasons. However, historical precedence does not guarantee that a particular species adequately represents a taxonomic class or an environment for a broad-based comparative toxicogenomics effort. In some cases, the choices of species for toxicological testing and the availability of genomics resources are concordant. For example, the genome-enabled mouse model will continue to serve as a valuable surrogate species for comparative toxicogenomics. In other cases, different regulatory agencies use a variety of different test species, and adding genomics resources to the calculus only complicates the selection and prioritization process. In addition to the alignment of regulatory use and genomics data, the representative quality of data from various species must be considered when selecting new toxicologic test species.

2.2   BRIEF REVIEW ON STUDIES USING COMPARATIVE GENOMICS

Comparative genomics has already yielded information relevant to better understanding of human disease and environmental susceptibility to toxicants and natural stressors, especially from mammalian species and aquatic organisms. The use of rodents as surrogate species to understand human disease is well established. The continuing editing and annotation of the mouse and rat genomes is enabling a deeper understanding of human disease and improving our understanding of the health effects of environmental toxins. This knowledge will undoubtedly lead to better risk assessments and provide the scientific foundation for good environmental policy. Additional surrogate species promise broader views of susceptibility and health outcomes, and they are essential to understanding the full extent of exposures to toxicants and stressors in mammalian and environmental species.

There is also a strong precedent for comparative genomics studies using fish. The completed sequencing of the pufferfish (*Fugu rubripes*) and zebrafish (*Danio rerio*) has already resulted in the identification of novel genes that appear relevant to understanding mechanisms of human disease. For example, the Fugu genome is about one eighth the size of the human genome, yet contains a comparable set of protein-coding genes (Aparicio et al. 2002). Close evaluation of the coding sequence suggests that more than 75% of the human genes are represented and easily identifiable. However, about 25% of human genes have no apparent homologues in Fugu. More interestingly, there were roughly 1000 genes discovered in Fugu with high sequence similarity with the human genome that had escaped previous identification (Aparicio et al. 2002). Thus, having the Fugu genome sequence has already helped improve annotation of the human genome.

Several fish species have been used to leverage this commonality, with knockout and transgenic models helping to determine gene function in humans. More than 500 zebrafish mutants that relate to human disease have been produced. In addition, zebrafish are routinely used to study development, carcinogenesis, and environmental health. They are the model of choice for testing endocrine disruption toxicants in

Europe. In the United States, zebrafish projects are widely supported as models for human disease because they are easier to manipulate than mammalian models.

If we have 2 fish models with fully sequenced genomes, do we need others for a comparative toxicogenomics research effort? The answer is yes, for 2 reasons. First, zebrafish and Fugu do not broadly represent the range of fish responses to chemical toxicants or natural stressors. Second, these models do not serve well as surrogate species for environmental risk assessment because they are not widely distributed and easily sampled in natural environments. Broad and easy access to natural populations is shown to be important by a recent study by Oleksiak et al. (2002), which demonstrated that the response of killifish to natural stressors is clearly complicated by genetic polymorphisms within natural populations. Their observations suggest that concentrated research efforts are required to extrapolate laboratory results from (often) inbred isolates to population-level effects due to genetic variability in the responses to toxicants and to natural stressors. Thus, access to genomics information from natural populations is critical to applying comparative toxicogenomics knowledge successfully.

Many environmental species have been developed to characterize toxicologic pathways activated by exposure to polyaromatic hydrocarbons, polychlorinated biphenyls, metals, pesticides, etc. The most popular models include frogs (*Xenopus laevis, Rana catesbeiana, Rana pipiens*), birds (*Coturnix*), fish (*Onchorinchus mykiss, Fundulus heteroclitus, Micropterus salmoides, Pimephales promelas, Cyprinodon variegatus*), and selected crustaceans (*Daphnia, Ceriodaphnia*). Toxicogenomics projects for most of these species are in progress in several laboratories. Initial data suggest that it will be possible to determine chemical class-specific gene expression patterns that are reflective of toxicology pathways and mode of action (Hamadeh et al. 2002; Larkin et al. 2003).

Caenorhabditis elegans and other invertebrate research models like *Drosophila melanogaster* also offer a comparative platform to understand human disease and environmental stressors. Their genomes are completely sequenced, and these systems have mature genetic toolkits. As a result, these classic model organisms have enabled multiple comparative studies to understand chemical or stress-induced gene expression changes and have enabled direct tests on the evolutionary preservation of gene function by using reverse genetics and reciprocal genetic transformations. Initially, *C. elegans* was developed as a surrogate for the nervous system (Brenner 1988) and it is a key surrogate species for Alzheimer's and Huntington's diseases. In addition, since its earlier uses, it has become invaluable for the elucidation of other biochemical pathways.

Efforts to evaluate *C. elegans* as a model for mammalian xenobiotic toxicity are under way. Anderson et al. (2004) found 5 compounds across 3 chemical classes used in behavioral toxicity tests that have an identical order of toxicity in *C. elegans* and mammals. Cole et al. (2004) followed with a study that found *C. elegans* could correctly predict the mammalian toxicity of 15 organophosphate insecticides. These researchers also found that the mechanism of action for this toxicity — inhibition of acetyl cholinesterase — was the same between the nematode and mammals.

Although research using invertebrates has many advantages over mammalian testing — including significantly less expense, ease of use, and greatly reduced time commitments — it also has limitations that need to be taken into account. For example, invertebrates may not possess some of the physiological processes found in mammals or higher animals, such as livers, kidneys, skin, and lungs. Even with these limitations, these simpler organisms have very similar cellular and genetic functions and, in some cases such as the nervous systems, share similar complex physiological processes with higher animals. They present many opportunities for use as a model for studying human disease and injury.

2.3 SELECTION CRITERIA FOR SURROGATE SPECIES

The preferred animal models will depend on the goals of the investigators. Each comparative study would employ model species that best fit the hypotheses to be tested. To begin addressing how one might select a set of models, we have divided research needs into 3 broad categories: 1) understanding core biological questions, including the function and evolution of genes involved in cellular processes, organ-specific physiology, development, and reproduction, among others; 2) understanding human disease; and 3) understanding the effects of toxic substances and natural stressors in the environment. For regulatory practices, genomics may eventually be able to distinguish broad classes of toxicants that activate common toxicological pathways. These studies should build on toxicological databases that have been assembled over dozens of years and that may offer the underpinning of phenotypic anchoring for studies. For environmental monitoring, such studies will require information not only at the gene level directed towards individuals, but also at the population level. Clearly, the list of surrogate species selected for these 3 paradigms will be different.

Well-defined selection criteria are required for choosing surrogate species that include the advantages and disadvantages of the model. The desired characteristics of an organism for molecular toxicological studies are similar to those common in model species for any biological investigation. These similarities include an understanding of the organism's basic biology, ease of handling and a short generation time, availability for use in the laboratory, access to natural populations, and the general ability to provide useful data for predicting outcomes of interest. For toxicogenomics research, many of the criteria used for choosing animal surrogates in the past are still applicable today. For example, it is still impractical to characterize the distribution of toxicant sensitivity across all species. Therefore, past strategies like those taken by the USEPA to identify and utilize the most sensitive keystone species may still be appropriate, with hopes of protecting ecological systems whose integrity depends on a large variety of untested species.

Certain organisms have been chosen in the past because they inhabit specific physical milieu, are broadly distributed and taxonomically important, or possess unique biological aspects (e.g., reproductive requirements). Yet, 3 selection criteria stand in contrast from past measures. First, the volume of toxicological information accumulating for select taxa within scientific and regulatory agencies' databases is a specifically important factor in the choice of surrogates for toxicogenomics study. A related consideration is the quality of these data. In some cases, we can ask

TABLE 2.1
General selection criteria for surrogate species

1) Globally distributed in the wild
2) All life stages tractable in the lab
3) Existing laboratory toxicological database
4) Ability to collect large numbers of organisms at specific life-stages in the lab (for statistical purposes)
5) Sample all life stages in the field
6) Readily available large number of samples from identifiable populations
7) Relatively short generation time
8) Existing field toxicological database
9) Known natural history
10) Model for human disease research

whether data for a particular species are predictive of relevant toxicological outcomes, what the diversity of environmental stressors assessed and their relevance to ecological settings or human health are, what is known about the comparative sensitivity of the species to environmental stressors, whether mechanistic studies have been performed and whether they indicate consistency with findings from other species, and whether data from toxicological studies have been validated between laboratories and, for ecological studies, between the laboratory and the field. Ideally, there should be some evidence that the organism can provide insight into toxicological concerns and lead to predictive outcomes.

Second, given the current costs and difficulties associated with developing genomics tools for research, the availability of an annotated genome is required. If genomics information is lacking, then basic information on the approximate size of the genome, number of chromosomes, and the ploidy level should be available to help formulate a plan to obtain genomics sequence. Third, because genomics tools and information are often narrowly applicable to divergent organisms, the phylogenetic positioning of taxa is an important criterion. Although no single species may be able to meet all selection criteria or provide useful data for all potential needs, Table 2.1 provides an unranked listing of 10 desired criteria for a surrogate species to be used in a toxicogenomics study.

2.3.1 Toxicologic Information

The process of selecting species for genomics investigations in toxicology should not overlook surrogate species currently employed for research. Toxicological databases are enormous. The scientific literature from 1965 to the present comprises more than 3 million toxicology-related records combining knowledge from the fields of biochemistry, physiology, pharmacology, and environmental sciences. To assess the relative merit of taxa based on their number of reported test results, we mined the ECOTOX database (http://www.epa.gov/ecotox/), which is divided into 2 sections: 1) the AQUIRE dataset contains information from toxicological tests using aquatic organisms, and 2) the TERRETOX dataset contains test results using terrestrial species.

Both datasets used here were last updated on 1 September, 2003. Although focusing on ecologically relevant chemicals restricts these data, they are good standards for ranking taxa by referencing 19,016 studies using 5022 species and totaling 455,194 records. The aquatic database exceeds the number of terrestrial data entries by referencing 15,124 studies compared to 3892. One possibility for this difference is that a lot of rodent data collected by pharmaceutical companies is not released into the public domain. For this reason, we present results for each section.

A figure that orders the top 25 aquatic and terrestrial metazoans listed in the databases according to their number of entries clearly demonstrates that most toxicological information is derived from a small number of taxa (Figure 2.1, appendix A). For instance, the top-ranked aquatic (rainbow trout) and terrestrial (mallard duck) species account for more than 10% and 13% of all data for animals in their respective databases. Furthermore, the 25 top-ranked organisms encompass more than 50% and 75% of all recorded information, respectively. These findings suggest that a modest number of species can be targeted for genomics studies while still capitalizing on the majority of accumulated toxicological knowledge. However, the array of biological diversity represented among these organisms is shallow. For example, among the 25 aquatic animals, 18 are fish, while all 3 freshwater invertebrates are cladoceran crustaceans commonly known as water-fleas (Appendix A; *Daphnia* and *Ceriodaphnia*). This paucity of information for large classes of animals is a weakness for comparative studies and is a common limitation among most databases.

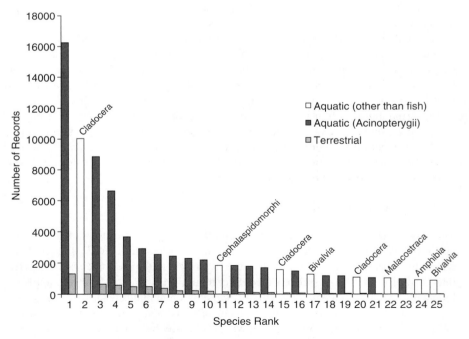

FIGURE 2.1 Histogram of ranked aquatic and terrestrial animal species based on the respective number of records for those species in USEPA ECOTOX toxicological databases. Species designations are listed in Appendix A.

In addition to the universal need for phylogenetic diversity within a list of proposed surrogates, our list should prioritize taxa from among the top models in ecotoxicology. In doing so, valuable information from past experimental and field studies can guide functional genomics investigations. In turn, discoveries that connect genome structure, gene expression, and animal responses to environmental toxicants can be linked more easily to individual fitness and population-level responses. Ideally, selected taxa should represent animals that populate a variety of key habitats (soil, marine, benthic freshwater, limnetic freshwater, etc.) for comparative ecological study. However, a second and third goal of comparative toxicogenomics is to select taxa with important features that refer specifically to human health and to basic physiological and cellular responses to toxicants. As a result, our toxicological information criterion, which is based solely on information within the ECOTOX database, is applied predominantly to select ecological surrogates and plays a lesser role in selecting surrogates for extrapolating toxicological genomics findings to humans.

2.3.1.1   Weighting Selection Criteria Based on Three Research Needs

The weight given to the criteria listed in Table 2.1 will vary depending on the type of study performed and on the research goals. For species used as models for studying core biological properties, the fewest criteria are required. Table 2.2 lists the criteria most relevant to this goal and highlights requirements for a short generation time and for maintaining large numbers of individuals in a laboratory during all stages of their development. However, for a species to provide data for human health assessments, it should have a similar physiological response to toxins and share target tissues with humans.

Finally, for an organism to serve as an ecological surrogate species, it requires the greatest number of characteristics (Table 2.2). First, the organism should have

TABLE 2.2
Selection criteria for surrogate species by area of extrapolation

Core biological function	Human health risk assessment	Ecological health risk assessment
All life stages tractable in the lab	All life stages tractable in the lab	Globally distributed in the wild
Existing laboratory-based toxicological database	Existing laboratory-based toxicological database	Existing laboratory-based toxicological database
Ability to collect large numbers of organisms at specific life stages in the lab	Ability to collect large numbers of organisms at specific life stages in the lab	Possible to sample all life stages in the field
Short generation time	Short generation time	Readily available large number of samples from identifiable populations
	Model for human diseases	Existing field toxicological database
		Known natural history

ecological relevance, including a broad distribution with readily accessible populations that can easily be cultured in the laboratory during all life stages. The culturing techniques should be standardized, and reference toxicity data should be available to determine their health and usability. Other desired traits include simple methods for synchronizing their development in order to provide consistency between exposure outcomes. Additionally, the species should occupy a strategic phylogenic position that allows for extrapolation to other species within the phyla.

2.3.1.2   Genome Information

Spurred primarily by the promise of accelerated discoveries in human health and pharmaceutical research, genome sequencing projects have grown to include a variety of animal species that permit comparative analyses of genome structures, gene functions, and interactions. As of August 2004, there were 10 completed genome sequencing projects and 71 projects at various stages of completion (Appendix B). The majority of these projects focus on mammalian taxa that intimately inform on human biology (primates) or are a currently major biomedical research model. Nonmammalian vertebrates with genome sequences — such as the South American opossum, the domestic chicken, the African clawed frog, and the zebrafish — offer important evolutionary insights on basic life processes (e.g., embryology, immunology, neurology), whereas more ancestral species at the root of the vertebrate phylogeny have profound implications for understanding developmental genetics. Other well-represented groups with cataloged genomes include close relatives of the premier genetic research models: *Drosophila* and *Caenorhabditis*.

   Altogether, these species cover a large range of animal diversity (Figure 2.2) and improve understanding of the molecular mechanisms underlying toxic responses to environmental contaminants. Investigations are currently under way to discover how findings obtained using these model species can be extrapolated to humans. However, there are many omissions from an ideal representation of animal diversity, especially from classes whose members are considered keystone species in ecological settings and whose attributes in natural environments are well documented. With the exception of some fish (tilapia, stickleback, salmon, trout) and fly (*Aedes*) species, there are no other freshwater animals with ecological standings whose genome is being sequenced, except for a branchiopod crustacean (*Daphnia pulex*). Although *Xenopus* are excellent amphibian models for genomics research and have an important toxicological database, their geographical distribution and population biology are restricted. Nematodes are also limited by their ill-defined population structures and cryptic early life stages in the wild, even though these species are presently the only soil-dwelling invertebrates with genomics and toxicological data.

2.4   SELECTION OF SURROGATE SPECIES

Currently, whole genome sequencing projects are focused on human health and the study of core biological processes. However, the current list of sequencing projects is lacking species that are the most appropriate ecological surrogates. Criteria for selecting these surrogates must rely first on the usefulness of the model for ecological

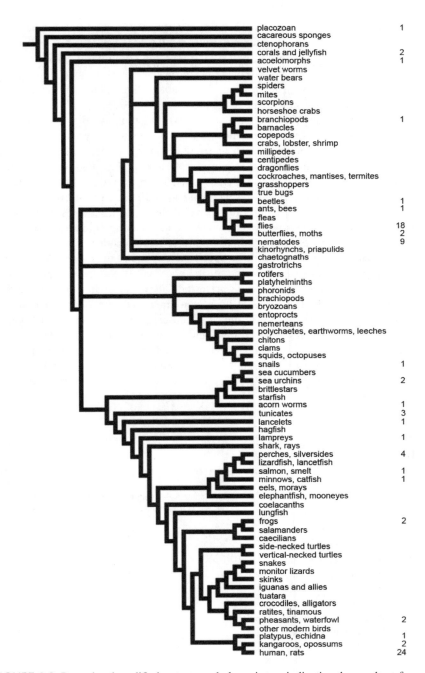

FIGURE 2.2 Pruned and modified metazoan phylogenic tree indicating the number of species from various classes with whole genome sequencing projects as of August 2004. (From Pennisi E. 2003. Science 300:1692–1697.)

studies and risk assessment and its sensitivity to toxicants. Secondary considerations are the niche the species hold in the environment, the organism's genome size, and the development of a genomics toolkit by the research community, including large insert genomics libraries, cDNA sequencing projects, a genetic map, microarrays for functional investigations, and a genomics database. Species that fit these criteria should be considered for new genome projects.

To select surrogate species for each area of extrapolation, we first made a partial list of current species (Appendix C) used for toxicological analyses. Each species was given a score based on the 10 criteria listed in Table 2.1. Depending on the area of extrapolation, different criteria were considered: core biological processes (criteria 2–4, and 7); human health (criteria 2–4, 7, and 10), and ecological health (criteria 1, 3, 5, 6, 8, and 9). Each criterion was weighted according to the relative importance of that item (Table 2.3) and a sum score was calculated for each species. The animals were then ranked. We subsequently chose the highest ranking animal of a particular class. If there was more than one choice, we investigated whether there was genome or EST sequence information for those species and whether they were already used for regulatory purposes or had a standard method for toxicology testing. Animals with the most positive outcomes were preferred. These selected species were then tallied onto the final chart (Table 2.4). The Venn diagram (Figure 2.3) illustrates the extent of overlap of these species in the 3 core research areas. A total of 17 species appeared as top contenders in all 3 areas and are listed in Table 2.5. The relative importance of the weighting scheme (Table 2.3) is admittedly biased by the research experience of the authors. However, our model scheme can be applied to other groups interested in developing lists of surrogate species for toxicogenomics or other "omics" studies.

TABLE 2.3
Criteria weighting for selection of surrogate species

Criteria	Core biological domain issues	Human health	Ecological issues
1) Widely distributed			3
2) All life stages tractable in the lab	2	2	
3) Existing toxicological database	1	3	3
4) Ability to collect large numbers of organisms at specific life stages in the lab	3	2	
5) Sample all life stages in the field			3
6) Readily available large number of samples from identifiable populations			3
7) Short generation time	3	3	
8) Existing field toxicological database			3
9) Known natural history			3
10) Model for human diseases		3	

TABLE 2.4
Ranking of animals from highest to lowest for each area of extrapolation

Group	Name	Habitat	Sum	Regulatory use	Standardized method	Genome sequence	ECOTOX ranking
	A. Species selected as surrogates for core biological processes						
Nematode	Caenorhabditis elegans	s	9		+	+	18
Mammals	Rattus norvegicus	t	9	+	+	+	4
Mammals	Mus musculus	t	9	+	+	+	6
Fish	Danio rerio	fw	9	+	+	+	14
Fish	Oryzias latipes/Pimephales promelas[a]	fw	9	+	+	+	6/3
Birds	Gallus gallus	t	9	+	+	+	9
Fish	Cyprinodon variegatus/Fundulus heteroclitus[a]	m	9		+		12/–
Birds	Sturnus vulgaris	t	9	+	+		
Birds	Coturnix japonicus/Colinus virginianus[a]	t	9	+	+		7/2
Birds	Falco sparverius	t	9	+	+		
Annelid	Eisenia fetida	s	9	+	+		3
Crustacean	Americamysis bahia	m	9				
Crustacean	Ceriodaphnia dubia	fw	9				15
Amphibian	Rana spp.[b]	fw/t	9		+		
Amphibian	Bufo bufo	fw/t	9		+		
Amphibian	Ambystoma tigrinum	fw	9				
Amphibian	Xenopus laevis	fw	9		+	+	
Other invertebrate	Dictyostelium spp.	t	8			+	
Insect	Apis mellifera	t	8			+	
Insect	Drosophila melanogaster	t	8	+		+	5
Amphibian	Xenopus (Silurana) tropicalis	fw	6		+	+	
Crustacean	Daphnia magna/pulex[a]	fw	6	+/–	+/–	–/+	2/20
Fish	Salmo salar	fw	6	+	+	+	21

TABLE 2.4 (continued)
Ranking of animals from highest to lowest for each area of extrapolation

Group	Name	Habitat	Sum	Regulatory use	Standardized method	Genome sequence	Ecotox ranking
Fish	Oncorhynchus mykiss	fw	6	+	+	?	1
Birds	Taeniopygia guttata	t	6	?	+		6
Birds	Anas platyrhynchos	t	6	+	+		1
Crustacean	Hyalella azteca	fws	6	+	+		
B. Species selected as surrogates for human health risk assessment							
Fish	Danio rerio	fw	13	+	+	+	14
Mammals	Mus musculus	t	13	+	+	+	6
Fish	Oryzias latipes/Pimephales promelas[a]	fw	13	+	+	+	6/3
Mammals	Rattus norvegicus	t	13	+	+	+	4
Nematode	Caenorhabditis elegans	s	13		+	+	18
Amphibian	Xenopus laevis	fw	12	+	+	+	
Birds	Gallus gallus	t	12		+	+	9
Birds	Coturnix japonicus/Colinus virginianus[a]	t	11	+	+	+	7/2
Amphibian	Rana spp.	fw/t	11		+		
Amphibian	Bufo bufo	fw/t	11		+		
Amphibian	Ambystoma tigrinum	fw	11				
Birds	Sturnus vulgaris	t	10	+	+		
Fish	Cyprinodon variegatus/Fundulus heteroclitus[a]	m	10		+		12/–
Crustacean	Americamysis bahia	m	10				
Annelid	Eisenia fetida	s	10	+	+		3
Crustacean	Ceriodaphnia dubia	fw	10	+			15
Birds	Sturnus vulgaris	t	10	+	+		
Insect	Drosophila melanogaster	t	10	+		+	

C. Species selected as surrogates for ecotoxicology and risk assessment

Group	Species	Habitat	Sum	Method	Regulatory use	Genome sequence	EXOTOX ranking
Amphibian	*Rana* spp.[b]	fw/t	18			+	1
Birds	*Anas platyrhynchos*	t	18	+	+	+	
Birds	*Falco sparverius*	t	18	+	+	+	
Crustacean	*Daphnia pulex/magna*[a]	fw	18/15	-/+	-/+	+/-	20/2
Fish	*Gasterosteus aculeatus*	fw, m	18			+	3
Annelid	*Eisenia fetida*	s	15	+	+	+	
Birds	*Sturnus vulgaris*	t	15	+	+	+	
Birds	*Colinus virginianus/Coturnix japonicus*[a]	t	15/12	+	+	+	2/7
Crustacean	*Americamysis bahia*	m	15	+		+	
Crustacean	*Ceriodaphnia dubia*	fw	15			+	15
Fish	*Cyprinodon variegatus/Fundulus heteroclitus*[a]	m	15	+		+	12/-
Fish	*Oncorhynchus mykiss*	fw	15	+		+	1
Fish	*Pimephales promelas/Oryzias latipes*[a]	fw	15	+		+	3/6
Fish	*Salmo salar*	fw/m	15	+		+	21
Insect	*Apis mellifera*	t	15			+	5
Mammals	*Mus musculus*	t	15	+		+	6
Amphibian	*Bufo bufo*	fw/t	15	+		+	
Amphibian	*Ambystoma tigrinum*	fw	12			+	
Crustacean	*Hyalella azteca*	fws	12	+		+	

[a] More than one species was identified as an important indicator in similar ecological niches.

[b] This includes *Rana catesbeiana*, *R. pipiens*, *R. rugosa*, and *R. temporaria*. See text for details.

Notes: Sum = the sum of weighted criteria for each area of extrapolation. Regulatory use = species that are currently used for risk assessment. Standardized method = method has been or is being developed by regulatory agencies. Genome sequence = availability of genome sequence information. EXOTOX ranking = rank in Appendix A.

t = terrestrial; fw = freshwater; fws = freshwater; m = marine; ms = marine sediment; s = soil; fw/t = part of life cycle as an independent organism spent completely in freshwater and part of life cycle as an independent organism spent on land; fw/m = part of life cycle as an independent organism spent in freshwater and another part spent as independent organism in a marine environment.

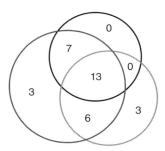

FIGURE 2.3 Venn diagram showing the number of surrogate species selected in Table 2.5 by the authors for comparative toxicogenomics, based upon 3 different evaluative paradigms: core biology, human health, and ecological health risk assessments. The 13 species encompassed by all 3 areas of extrapolation are listed in Table 2.5.

TABLE 2.5
Species selected as surrogates for core biological function, human health, and ecotoxicology

Scientific name	Common name
Ambystoma tigrinum	Tiger salamander
Americamysis bahia	Mysid shrimp
Bufo bufo	Toad
Ciona intestinales	Sea squirt
Coturnix japonicus/Colinus virginianus[a]	Japanese/bobwhite quail
Cyprinodon variegatus/Fundulus heteroclitus[a]	Sheepshead minnow/fundulus
Ceriodaphnia dubia	Water-flea
Eisenia fetida	Earthworm
Falco sparverius	American kestrel
Mus musculus/Rattus norvegicus[a]	Mouse/rat
Oryzias latipes/Pimephales promelas/Danio rerio[a]	Medaka/fathead minnow/zebrafish
Rana spp.[a]	Frog
Sturnus vulgaris	Starling

[a] More than one species was identified as important indicators in similar ecological niches.

2.5 DISCUSSION

We have considered a wide range of animals that could act as surrogates for a toxicogenomics research program. However, the careful reader will identify obvious gaps in the taxonomic representation of selected species. For instance, there is limited deliberation in this review for whole classes of familiar animals (e.g., reptiles, numbering more than 8000 described species); to a certain degree, there is only token consideration for members within certain subphyla (e.g., hexapods, numbering in the millions) not to mention the absence of whole phyla. The following discussion of various groups reflects our particular knowledge and fields of expertise. The absence of any group of animals should not be interpreted as having no endorsement. Yet, based on our method for assessing the merits of groups that were evaluated, we expect that species with endorsements will significantly advance the field of comparative toxicogenomics.

2.5.1 MAMMALIAN MODELS

Great amounts of effort and attention have been placed on the development and application of mammalian genomics to toxicology, primarily as it applies to human health research and risk assessment. Although human biology and toxicology are ideally studied using human tissues and derivative cell lines, in many cases this approach is neither feasible nor ethical. The vast majority of current toxicogenomics research is conducted using mice and rats, and these rodent models have generally served toxicology very well for identifying hazard and characterizing risk to human health. Continued technological progress allowing for miniaturization, increased sensitivity, noninvasive imaging, and use of alternative in vitro or in silico approaches has increased the potential for using human models for molecular epidemiology and laboratory-based studies. This humanization of toxicogenomics promises to reduce and perhaps ultimately replace the present dependence on animal testing.

Additional and important toxicological data are also derived from rabbit, dog, and nonhuman primate studies, but genomics studies with these species have been more limited due to increased costs and decreased genomics information. Thus, the present status of using mice and rats for the majority of toxicogenomics efforts is still prudent and fruitful. The consideration of mammalian models, beyond rodents and humans, for ecotoxicogenomics is problematic due to economic and ethical considerations. Nonetheless, the foundations for just such an approach are being laid with the genome sequencing and annotation of 28 mammalian species beyond human, mouse, and rat (http://www.genome.gov/10002154).

2.5.1.1 Core Biological Studies and Human Health

By applying the selection criteria described in this chapter and factoring in considerations of economic and societal priorities, it seems reasonable to maintain the focus of mammalian toxicogenomics on 2 species: *Mus musculus* and *Rattus norvegicus*. These species are globally distributed, have a wide range of life-stage models in the laboratory and field, have enormous quantities of existing toxicological and genomics data, and provide data that can be extrapolated to human health risk assessments. The great advantage held by mice and rats over humans is their utility and relative economy in controlled laboratory in vivo studies.

Rattus norvegicus, the brown Norway rat, was originally limited to China but now has worldwide distribution as a human commensal (Armitage 2004). Though originally native to Asian forests and brushy areas, today *R. norvegicus* prefers habitats alongside human populations. Nearly every port city in the world has a substantial population of these rodents; they occupy a variety of habitats including garbage dumps, sewers, open fields and woodlands, basements, and nearly anywhere else that food and shelter might be found. The massive reproductive capacity of Norway rats and their relatively large size make them the dominant rat species in many locales. Their short gestation period (21–25 days) makes them an ideal laboratory species from the viewpoint of efficiency. A significant fraction of laboratory toxicology worldwide has been and is conducted using *R. norvegicus*. This is especially true for reproductive toxicology, for which the rat is the preferred species (USEPA 2004). Recent advances in rat genome sequence and annotation have created a wealth of rat genome resources (NCBI 2004; Rat Genome Database 2004; Genome Sequencing Project Consortium 2004). These resources include genomics tools and technologies that have been applied by toxicologists to understand environment–gene interactions.

Mus musculus, the house mouse, was originally distributed from the Mediterranean to China, but now exists worldwide as a human commensal (Ballenger 1999). *M. musculus* generally lives in close association with humans in houses, barns, granaries, etc. *M. musculus* also occupies cultivated fields, fencerows, and wooded areas, but seldom strays far from buildings. Because of their association with humans, *M. musculus* are able to inhabit inhospitable areas that they would not be able to occupy independently. Similarly to the rat, *M. musculus* is characterized by tremendous reproductive potential. Various strains have been developed by mouse fanciers and for laboratory use. Because of their smaller size, mice are more economical than rats for laboratory research. Furthermore, despite recent advances in rat genomics, mice still hold an advantage in genetic information and resources, in part due to the large numbers of mutant and transgenic strains available to researchers (International Mouse Strain Resource 2004).

Disadvantages of using mice include the fact that, historically, the rat has been used most often to identify environmental chemical hazards, especially developmental and reproductive hazards (Gray et al. 2004) and this trend is likely to continue. Furthermore, compared with rats, mice have a higher incidence of spontaneous malformations and difficulty maintaining pregnancies under testing conditions (Jahnke et al. 2004). Finally, the larger rat is an easier source for tissue and biospecimens, including multiple blood samples for clinical chemistry, hematology, and serology (e.g., hormone measurements).

The 2 rodent species (mouse, rat) are complementary to one another as models for human biology, disease, and toxicology. Physiological and genomics comparisons (Genome Sequencing Project Consortium 2004) typify the similarities and differences across the 3 species. The best solution for comparative toxicogenomics is to move ahead with projects comparing both rodent species to humans to generate information about gene–environment interactions significant to human health.

2.5.1.2 Ecotoxicology and Risk Assessment

Consideration of appropriate mammalian models for ecological risk assessment certainly should include consideration of species beyond *M. musculus* and *R. norvegicus*. There are currently 31 mammalian species for which there are active genome sequencing projects (http://www.genome.gov/10002154). Rankings of terrestrial animals based on the number of records in ecotoxicological databases identified 3 mammalian species: #4, *R. norvegicus*, #6, *M. musculus*, and #11, the North American deer mouse *Peromyscus maniculatus* (Appendix A). This result reinforces the current selection of *M. musculus* and *R. norvegicus* as species for comparative toxicogenomics. The worldwide range of these rodent species further supports this selection and indicates the potential for their use in ecotoxicology.

2.5.2 Aquatic Models for Human and Ecological Health

2.5.2.1 Core Biology and Human Health Models

The top selected species in our list for fish for core biology and human health were zebrafish and medaka. This is primarily because they have their genomes almost completely sequenced and annotated and have already been used for numerous studies. For example, a recent study by Amsterdam et al. (2004) used insertional mutagenesis to obtain 315 genes out of an estimated 1400 genes required for zebrafish embryonal and early larval development. The article reports that 72% of the isolated genes had homologues in yeast, 93% in invertebrates, and 99% in humans — suggesting that this set of genes will be invaluable in identifying genes of interest to development in other species. Because the genomes for yeast, fly, and nematode are already sequenced and available for comparisons, the authors were able to identify the set of genes common to all, the set that was animal specific, and the set that was vertebrate specific. These types of multispecies comparisons are very strong in pinpointing core biological processes, which are conserved among all life forms and thus can give insight into how basic biochemical pathways are regulated.

The availability of morpholinos and RNAi technologies to knock down or disrupt specific genes has made zebrafish an invaluable model to determine the function of specific genes in a whole animal context. Finally, the introduction of transgenes into the zebrafish has also worked in a prospective way to determine the roles genes may have during development. These approaches, combined with comparative "omics" technologies, are powerful and portend an exciting systematic analysis for systems involved in core biology. Moreover, the *Fugu* genome project has allowed identification of genes in humans and higher vertebrates and it is a powerful model for core biology. *Fugu* is considered a model of equivalent value with zebrafish and medaka, although it does not appear in our final list because it has not been used extensively for toxicologic studies.

In addition to serving as models of choice for core biology, zebrafish and medaka are excellent resources for the study of human disease. For example, zebrafish has been used as a model for studying pyruvate dehydrogenase defects, which lead to severe neurological dysfunction, growth retardation, and death (Taylor et al. 2004). An entire volume of *Mechanisms of Development* (volume 121, 2004) was devoted to the uses of medaka in core biology and human disease. Sakamoto et al. (2004) made use of the highly conserved synteny between medaka and *Fugu* to identify a gene that may be involved in hypochromic anemia. They suggest that mutations in the gene for δ-aminolevulinic acid dehydratase may play a role. Numerous other papers have used this approach to begin to identify potential genes involved in human disease (Keller and Murtha 2004; Lambrechts and Carmeliet 2004; McMahon et al. 2004; Schwartz et al. 2004; Berghmans et al. 2005; Drummond 2005; Epstein and Epstein 2005; Hill et al. 2005).

Other fish species not listed in our tables have also been used to study human disease. The little skate (*Raja erinacea*) is an exceptional model for intrahepatic cholestasis, a liver disease (Cai et al. 2001). Furthermore, certain fish species serve as models for specific human diseases; for example, swordfish interspecies hybrids (*Xiphophorus maculatus* and *X. helleri*) are excellent models for UV-induced melanoma (Nairn et al. 2001) and the dogfish shark (*Squalus acanthias*) for sodium and chloride transporters involved in kidney disease (Forrest 1996). These aquatic models have been developed specifically for the diseases they represent rather than as global models of human disease. Thus, at the present time they are considered secondarily for genome sequencing resources.

2.5.2.2 Ecotoxicology and Risk Assessment

Environmental considerations are quite different from core biology and human health. Seven fish models were short listed in Table 2.4: fathead minnow (*Pimephales promelas*), fundulus (*Fundulus heteroclitus*), medaka (*Ozyrias latipes*), rainbow trout (*Oncorhynchus mykiss*), salmon (*Salmo salar*), sheepshead minnow (*Cyprinodon variegatus*), and the 3-spined stickleback (*Gasterostreus aculeatus*). These fish occupy different niches in the environment and were chosen to represent these habitats as discussed next.

Aquatic models for freshwater. For comparative field studies, models are needed for risk assessment paradigms based on fish that are commonly found in the environment and subject to exposure to natural and chemical stressors. In many instances, the more developed laboratory models will not work for ecological risk assessment because they are not commonly found in the environment and may not respond to the same concentrations of toxicants. Three species from Table 2.4 represent species that inhabit large regions of fresh bodies of water, including the fathead minnow, medaka, and rainbow trout. Fathead minnow and medaka are small fish that can be found in the wild in North America and Japan, respectively. Together with zebrafish, they have been chosen as models for endocrine disruptor research in the United States, Japan, and Europe. All 3 species are fractional spawners that can easily be reared and tested in the laboratory setting.

For many toxicologic studies, these 3 fish are viewed as similar with respect to their general responses to toxicants, even though they vary in their sensitivities to low-dose exposures. Rainbow trout is also on the list of candidate surrogate species because it is found naturally in cold water in North America and it has an annual synchronized reproductive cycle, typical of many larger wild fish. It is at the top of the food chain and bioaccumulates lipid-soluble contaminants, thus making it more sensitive to aquatic contamination. Because it is highly valued in the sports fishery arena, it is a direct link with human health. While the models described here are best suited for North America, it is important to note that they may not be the best for assessing risk in other parts of the world and that species that are common and threatened in those environments must also be developed.

Based on the number of citations in the ECOTOX database, rainbow trout is the number one model for environmental toxicology in the United States (Appendix A). Fathead minnow follows in the number-3 position. A substantial database of toxicological endpoints and physiological outcomes for both species has been developed. Their life histories are well studied and documented, and many protocols have been standardized for their use in risk assessment. Thus, it is easy to make connections among toxic effects, adverse physiologic endpoints, and changes in gene expression. Studies that pull these 3 areas together will enable validation of the use of omics technologies in risk assessment.

Many genomics resources are developed for the rainbow trout. A bacterial artificial chromosome (BAC) library of about 185,000 clones has been developed and is currently being sequenced and probed for homologous genes (Palti et al. 2004). Several other omics tools are being developed for its integration into comparative studies (Thorgaard et al. 2002). In Europe and Japan, extensive use of the zebrafish and medaka for environmental work and risk assessment make these equivalent models for ecotoxicology work in those parts of the world. But neither model has been accepted for risk assessment in the United States. In collaboration with the USEPA, the Joint Genome Institute has sequenced several fathead minnow cDNA libraries for a total of 250,000 sequences that are now available in the EST databases. This information is excellent but incomplete. Our recommendation is to develop a genome project for the fathead minnow.

Aquatic models for estuarine waters. For estuarine work in North America, 2 models have been used extensively: the sheepshead minnow and fundulus. As with the freshwater models, extensive databases exist documenting the effects of toxicants and natural stressors on these models. For fundulus, an in-depth study using cDNA microarrays was used in a landmark study to determine natural variability of the fish in the ecosystem (Oleksiak et al. 2002). The sheepshead minnow model was also exploited to develop gene expression fingerprints for exposure to estrogen and estrogen mimics in a study that showed that chemical class-specific gene expression fingerprints could be obtained for these types of compounds (Larkin et al. 2003). Sheepshead minnow has been developed as a testing model by USEPA and is an accepted model for risk assessment. Both models are excellent for reporting on this environmental niche and are thought of as equivalent models. There is substantial cDNA information for fundulus and a much smaller cDNA database for sheepshead minnow. The recommendation is to further develop these models as EST libraries and to choose the fundulus model for full genome sequencing to allow for ample comparative toxicogenomics work in this ecosystem.

Aquatic models for marine waters. The Atlantic salmon is an anadromous fish that is hatched in freshwater, spends most of its adult life in the sea, and then returns to freshwater to spawn. A large genomics project is under way in Canada (GRASP; Genomics Research for Atlantic Salmon Project) (Rise et al. 2004). This project brings together researchers from industry, academic institutions, and government to study this important fish. Various hormonal circuits are involved in the transition from freshwater to salt water and back, some of which can be targeted by environmental contaminants. Early results from the genome project hold promise in that this model may work for other salmonids, including the Pacific variety and rainbow trout. It also is a valuable model for genome studies because its genome recently underwent duplication and may give answers to fundamental questions regarding gene duplication, evolution, and adaptation.

The 3-spined stickleback is also of considerable interest. This group of fish has recently undergone a radial evolution, mainly in freshwater systems and in some marine environments, and they are found over the entire world. Many researchers are interested in their evolution, physiology, and ecology, making this a good model for the study of evolution (Schluter 2004). This model has also surfaced as an important model for endocrine disruption because of its response to androgenic substances (Katsiadaki et al. 2002). While sticklebacks are not currently targeted for risk assessment, their presence in natural waters (fresh and marine) around the world makes their use important. Currently, a BAC library is being developed for stickleback research (Stanford Genome Evolution Center) and it is certain that other omics tools will soon follow.

Other models for environmental toxicology have not been included on our list — not because they are not important but because their use in toxicology and risk assessment is specific to their environments and they are not distributed worldwide. Thus, with limited funds to be devoted to developing omics tools, they remain as second-tier models, even though some laboratories across the world have engaged in genome projects. Included in this group are flounder (Sheader et al. 2004), carp (Gracey et al. 2004), and tilapia (Katagiri et al. 2001), all of which currently have BAC libraries and sequencing projects.

2.5.3 Amphibian Models

Amphibia have long been regarded as sentinel species of the environment. There are currently 5797 species of amphibia worldwide: 5083 Anura (frogs and toads), 544 Caudata (newts and salamanders), and 170 Gymnophiona (caecilians) (Amphibia-Web 2005). Many of these species were identified only within the last 20 years. Three amphibian groups were selected as surrogates for core biological function, human health, and exotoxicology models: *Ambystoma tigrinum* (tiger salamander), *Bufo bufo* (common toad), and several Ranid frog species (Table 2.5).

2.5.3.1 Core Biology and Human Health Models

Several species of amphibia serve as important models for understanding basic and comparative cell or developmental biology, physiology, endocrinology, cancer biology, neurology, immunology, biochemistry, and toxicology. The most commonly used frog species for laboratory-based studies is *Xenopus laevis*, from the Pipidae family. It is an unusual frog in that it does not possess a tongue, and the adult stage is completely aquatic. However, its rapid development, well-established husbandry, and ease of experimental manipulation have made it a mainstay model for cellular and developmental studies.

This species forms the basis of the frog embryo teratogenesis assay-Xenopus (FETAX) assay (Courchesne and Bantle 1985) and a metamorphosis assay for thyroid hormone axis disruption and a reproductive assay are actively being developed by Organization for Economic Collaboration and Development (OECD) member countries. The generation of transgenic lines is possible (Amaya and Kroll 1999) and this feature has been used extensively in many areas of research. A considerable amount of EST and cloned cDNA information is publicly available (>400,000 nucleotide sequences in Genbank) along with DNA arrays from commercial and academic sources. However, genomics efforts in this species are limited due to the allotetraploid nature of the genome.

This significant drawback as a genetic system has led to the development of another related species for genomics analysis, *Xenopus* (*Silurana*) *tropicalis* (West African frog). Although there still is controversy as to the classification of the genus, this species is diploid and is therefore better suited for genetic approaches including gene knockout studies. In addition, it has a relatively small genome size for a frog (1.7×10^9 base pairs), inbred lines are available, and it is amenable to multigenerational studies requiring only 4 months until sexual maturity. A genome sequencing effort with a BAC-based physical mapping will be completed in 2006 (http://genome.jgi-psf.org). Complementary EST sequencing projects using stage-specific libraries at several institutions worldwide have resulted in over a million nucleotide sequence entries in Genbank to date.

Salamanders are also important model organisms for biological and biomedical research. The most commonly used species are *Ambystoma mexicanum* (axolotl) and *Ambystoma tigrinum* (tiger salamander). These species are used in the study of tissue regeneration, vision, neural and renal biology, embryogenesis, heart development, and olfaction. The genomes of these species are 20 times larger than that of *X. tropicalis*, making sequencing and annotation of these genomes challenging. However, a concerted effort in sequencing EST libraries and genome mapping is under way (http://salamander.uky.edu/). More than 50,000 nucleotide sequences are currently available in Genbank.

2.5.3.2 Ecotoxicology and Risk Assessment

Amphibia species like *Xenopus laevis* serve as excellent laboratory species for ecotoxicological studies due to their known natural history and ease of breeding and maintenance in the lab. *X. tropicalis* does not have the same degree of scientific knowledge base and toxicological data as *X. laevis*. Nevertheless, with its genomics tools and relatively straightforward husbandry requirements, it is anticipated that this knowledge gap will quickly be closed in the future. Most importantly, it remains to be seen whether this species shows similar sensitivity to environmental contaminants as its sister species and how its sensitivity compares to that of other frog species.

Ambystoma and *Xenopus* have limited geographic distributions. *A. mexicanum* is only found in certain lakes in Mexico, whereas *A. tigrinum*'s range also includes North America (AmphibiaWeb 2005). Similarly, both *Xenopus* species are found in limited regions in Africa, although introduced *X. laevis* have established populations in the San Diego area (AmphibiaWeb 2005). There is a great need to gather genomics information from environmentally relevant species. Regulatory-based testing (deriving toxicological data for a new substance or compliance monitoring) must be relevant to the environment into which the substances or effluents are to be discharged. Indeed, the vast majority of toxicological studies are done on members of the Ranidae family, which is distantly related to Pipidae (Pough et al. 2001). Two other families to which several other species used in toxicological testing belong, Bufonidae (e.g., species *Bufo bufo*) and Hylidae, bear closest phylogenetic relationship with Ranidae.

Unlike the limited number of species and range of the Pipids, Ranids show the greatest worldwide distribution with over 700 species known. Known as the "true frogs," most Ranid species lay their eggs in the water and have feeding tadpoles and terrestrial adults. A few species have direct development, bypassing the free-living tadpole stage. Many countries around the world are moving in the direction of developing regulatory test methods using native species rather than solely relying on *Xenopus* in recognition of the need for addressing environmental relevance, which has an impact on toxicant response.

The most commonly used species for toxicological studies are *Rana pipiens* (Northern leopard frog), *Rana catesbeiana* (North American bullfrog), *Rana rugosa* (wrinkled frog), *Rana esculenta* (edible frog), and *Rana temporaria* (European common frog), although a wide range of species have been examined in the scientific literature. All of these species have shown sensitivities to environmental pollutants and developmental abnormalities in the wild; for some, regional populations are declining or threatened. Only one of these species, *R. catesbeiana*, is found on more than a single continent with populations in North and South America, Europe, and Asia. Incidentally, because of its large size, this species in particular is highly prized for human consumption (frog legs). These Ranid species, along with the European common toad, *Bufo bufo*, form the core group of amphibia of particular interest in the development of genomics tools. Over the last decade, close to 6000 publications on these species attest to their popularity (Web of Science 2005).

Two types of ecotoxicological assays are currently under development with *Ranid* species: a metamorphosis assay and a reproductive assay, in which toxico-genomics endpoints could be integrated. The former assay is much farther along in development and is capable of detecting a disruption in the thyroid axis because the metamorphosis of the larval tadpole is completely dependent upon thyroid hormone signaling. For this, several species are being used depending upon the country; these include *R. rugosa*, *R. pipiens*, and *R. catesbeiana*, among others.

A current challenge in these types of assays resides in their heavy reliance on morphological criteria (this holds true for the *Xenopus*-based assay mentioned previously). Morphological changes occur after many days to months, depending upon the developmental stage chosen for the assay. There is considerable variation in the normal development of these animals, even when starting with the same stage animals, and this difference becomes greater the longer the assay is run. A major target of thyroid hormone action is alteration of transcriptomes and proteomes, some of which occur in a much shorter time span prior to any overt morphological changes; therefore, the use of genomics tools would have a great impact in assay development provided that the link between transcriptome and proteome components and adverse effect is established. The same argument holds true for the reproductive assay.

In the laboratory setting, standardized methods are being developed for endocrine disruptor monitoring for *R. catesbeiana, R. pipiens* (North America), *R. temporaria* (Europe), and *R. rugosa* (Japan). Unlike *Xenopus*, no inbred lines for these species currently exist, but breeding programs are currently under way. The generation time for these animals is relatively short (~90 days to complete metamorphosis from the egg stage, 2 to 3 years until sexual maturity) with the exception of *R. catesbeiana*, which requires 1 to 3 years to complete metamorphosis from the egg stage and an additional 3 to 5 years to become sexually mature. For morphologically based assays, this presents a drawback compared to the faster developing species. However, this also represents an advantage in that this species is excellent for chronic exposure studies, has the highest metamorphic survivorship rate, and has a low and consistent baseline over a broader time window for thyroid hormone and sex differentiation studies. All *Rana* species are readily sampled from their natural populations and experiments using caged animals for in situ exposures are highly feasible to assess real-life exposures. This is simply not possible with *Xenopus*.

Certainly, access to genomics information from *Xenopus* has aided in the study and development of limited molecular tools for other frog species. For example, a cDNA array made with conserved ~500-bp open reading frame fragments derived from *Xenopus* sequences (Crump et al. 2002; Helbing et al. 2003) was successfully used to identify thyroid hormone responsive gene transcripts in *Rana catesbeiana* (Veldhoen et al. 2006). However, oligo-based and 3'-UTR-based detection methods (such as the Affymetrix *Xenopus* chip) have very limited use on Ranids, given that the extremely low degree of nucleotide sequence conservation negates specific hybridization signals. Open reading frame information is moderately applicable from *Xenopus* to highly conserved peptide fragments for proteomics analyses (e.g., peptide mass mapping), thus reducing the need to proceed to the more expensive peptide sequence determination using tandem mass spectrometry. Less conserved proteins (which often fall into the realm of the "most interesting proteins") cannot be dealt with in this way, however. Thus, true analysis of the proteomes is not possible without extensive information about the genome for the specific species of interest.

A major challenge in sequencing most frog genomes is the size. The average Ranid and Bufonid genome size is about 3 times larger than the genome of *X. tropicalis* (Gregory 2005). Efforts should be made to increase the gene sequence database for as many frog species as possible because of the diverse life histories, breeding strategies, natural ranges, etc. Like the approach taken with salamanders, this could be accomplished by generating EST libraries as well as targeted PCR (polymerase chain reaction) amplicons using degenerate primers based on sequence information from *Xenopus* and other species. The latter approach and more conventional cloning approaches have already yielded more than 2000 Ranid nucleotide sequences in Genbank.

2.5.4 *Ciona*

Ciona intestinalis (sea squirt) is a nonverterbate chordate that has long been used as a model for studying the evolutionary origins of the chordate lineage. It has a small genome (2×10^8 bp) and has been sequenced by the Joint Genome Institute (Dehal et al. 2002). It is an important marine representative that is available worldwide throughout the year. The life history is well known and cellular processes are easily observed. A cDNA microarray approach has been used on tributyltin-exposed animals (Azumi et al. 2004); to identify 200 genes that are strongly differentially expressed.

2.5.5 AVIAN MODELS

Several species of birds have been used as surrogate species. Many species can be held and bred in captivity, and all species are useful for studying embryonic development because development occurs in ovo rather than in utero. The effects of environmental contaminants on gene expression in embryos are readily studied by injecting chemicals of interest into the egg and/or by adding chemicals to primary cultures of various organs (e.g., liver, heart, brain). A brief summary of the characteristics of birds that have made them useful as surrogate species is presented next. Some species have been extremely useful for all 3 general characteristics of surrogates (basic biological process, human health, ecological health or ecotoxicology), but the use of others has been mainly limited to studies on ecological health or ecotoxicology. For the latter group, the quantity of DNA sequencing data presently available is extremely limited.

Chicken (Gallus gallus). The chicken is an important model for studies on gene regulation, embryology, and development, and it has also been used to test for the toxicity and mechanisms of action of environmental contaminants. It has a relatively short generation time (6 months) and it is relatively easy to conduct studies in ovo and in chicken primary cell cultures (e.g., liver, heart, brain). In March 2004, the National Human Genome Research Institute (NHGRI) announced that the first draft of the genome sequence of the red jungle fowl, the ancestor of the domestic chicken, was completed (http://www.nhgri.nih.gov/11510730). The chicken genome is the first avian genome to be sequenced, and the initial assembly, which is based on sevenfold sequence coverage, can be found in GeneBank (http://www.ncbi.nih.gov/Genbank).

Zebra finch (Taeniopygia guttata). A great deal of research has been carried out on the behavior, population biology, and ecology of songbirds, and the most studied species is the zebra finch (*Taeniopygia guttata*). The zebra finch has a short generation time (4 months), is relatively easy to breed in captivity, and has been used for ecotoxicology studies. A proposal for the construction of a BAC library of the zebra finch genome was submitted to the NHGRI in 2001, and a BAC library is available from the University of Arizona (http://www.genome.arizona.edu/orders).

Mallard duck (Anas platyrhynchos). The species *Anas platyrhynchos*, which includes the wild mallard, Pekin, and several other strains of ducks, has been used for a rather limited number of studies on basic biology (other than that specifically related to birds), but it has been used for ecotoxicology studies in the field and in the laboratory. Mallards and other strains of *Anas platyrhynchos* can be held captive and bred in the laboratory, with a generation time of approximately 12 months. On August 20, 2004, there were 751 DNA sequence data entries in PubMed. To our knowledge, a large genome sequencing project for *Anas platyrhynchos* has not been formally proposed to the NHGRI or to another agency responsible for large DNA sequencing projects.

Kestrel (Falco sparverius). A great deal of research has been carried out on the behavior, population biology, and ecology of the American kestrel. It has a relatively short generation time and can be held captive and bred in the laboratory. It has been used for ecotoxicology studies in the laboratory and in the field and is often used as a surrogate species for other birds of prey. On 20 August 2004, there were only 13 DNA sequence data entries in PubMed. To our knowledge, a large genome sequencing project for the American kestrel has not been formally proposed to the NHGRI or to another agency responsible for large DNA sequencing projects.

Herring gull (Larus argentatus). A considerable amount of research has been carried out on the behavior, population biology, and ecology of the herring gull. The generation time is 2 to 3 years, and it can be housed in the laboratory. However, there have been few laboratory studies with adult or juvenile birds. On 20 August 2004, there were 186 DNA sequence data entries in PubMed. To our knowledge, a large genome sequencing project for the herring gull has not been formally proposed to the NHGRI or to another agency responsible for large DNA sequencing projects.

European starling (Sturnus vulgaris). A considerable amount of research has been carried out on the behavior, population biology, and ecology of the European starling. The generation time of this species is approximately 12 months; it can be housed in captivity and has been used for ecotoxicology studies in the field and in the laboratory. The amount of gene sequence data for the European starling is extremely small (34 entries in PubMed on 20 August 2004).

Japanese quail (Coturnix corurnix japonica). The Japanese quail has been widely used in many areas of biomedical and behavioral research and has also been used extensively for toxicological studies on pesticides and other environmental contaminants. The species is easy to breed in the laboratory and has a short generation time of approximately 3 months. On 20 August 2004, there were 335 DNA sequence data entries in PubMed. To our knowledge, a large genome sequencing project for the Japanese quail has not been formally proposed to the NHGRI or to another agency responsible for large DNA sequencing projects.

2.5.5.1 Core Biological Studies and Human Health

The chicken has been used as an important model organism for many studies on gene regulation. For example, many of the pioneering studies of steroid hormone regulation were carried out in the chicken oviduct, and modern methods in molecular biology, biochemistry, and physiology with the chicken continue to provide important insight into complexity of gene regulation in vertebrates. The distinction between T- and B-cells was discovered in the chicken, and because chicken red blood cells contain a nucleus, they have been an important model system for studying chromatin structure. There are many other examples of how studies on the chicken have contributed to understanding basic biological processes. For example, the CCCTC-binding factor appears to play an important role in imprinting. According to a PubMed search conducted by McPherson and colleagues (http://www.genome.gov/Pages/Research/Sequencing/SeqProposals/Chicken_Genome.pdf), the user community for the chicken is larger than that for all other nonmammalian species, including zebrafish and Fugu.

The zebra finch is another useful model species for studying a variety of basic biological processes, including sexual differences in neuronal structure and function, the neuronal basis of learning, adult neuronal replacement, and steroid synthesis in the brain. A brief summary of some of the seminal findings can be found in http://www.genome.gov/Pages/Research/Sequencing/BACLibrary/zebraFinch.pdf. Several of these studies have had major impacts on the field of neurobiology because some the discoveries made first with zebra finches (and other songbirds) have been found to be generally true in mammals.

Chicken and zebra finch have been the major models used for human disease. For many years, the chicken has been a major model organism for studies of viruses, cancer, and heart disease. The first tumor virus was identified in the chicken, and avian leucosis viruses are among the most intensively studied retroviruses.

Molecular biological studies of zebra finch song learning have contributed to an understanding of the function of synuclein, which was subsequently discovered to be associated with Parkinson's disease in humans. The first compelling evidence that adult brains can generate new neurons was discovered in songbirds, and it is now well established that adult human brains can also undergo neurogenesis. An important goal of several laboratories is to develop methods to introduce stem cells into the human brain to replace or repair damaged neuronal circuits. Research on the mechanisms of neurogenesis in the zebra finch will continue to be important for understanding neuronal processes in humans.

The other avian models — mallard duck, American kestrel, herring gull, and European starling — have had a limited number of studies for biological processes or human health, but are extensively used for ecotoxicology and risk assessment (described in Chapter 5). There are some studies with the mallard duck on the mechanisms underlying osmoregulation in marine birds. Studies in this area for the American kestrel are limited to studies on vision and on the effects of electromagnetic fields from transmission lines on melatonin.

The European starling has been a useful species for a limited number of studies on the development of song. In this regard, the starling is of interest because it is less extreme in the degree of sexual dimorphism of the song system and song behavior than the zebra finch. The European startling has also been used to better understand the molecular mechanisms that help explain vision and auditory acuity in birds and other organisms. To our knowledge, no studies have used the European starling as a model for understanding mechanisms underlying human health and disease.

2.5.5.2 Ecotoxicology and Risk Assessment

The Japanese quail has been a commonly used as a model species for studying the effects of pesticides and other environmental contaminants in birds (particularly in the United States). While many of the endpoints have been based on overt toxicity, it would be expected that the availability of the genomics sequence for this species would help investigators make links between subtle effects on gene expression and overt toxic responses.

The chicken is a very useful model avian species for studying the effects of environmental contaminants in birds. Although large differences exist among avian species in sensitivity to most contaminants (e.g., dioxins, dioxin-like PCBs, organo-phosphate pesticides), laboratory studies with the chicken (in vivo and in cultured cells) are useful for environmental risk assessments and for understanding the mechanisms of action of toxicants. Although the chicken does not exist in the wild, in most areas of the world it has been used with success to monitor the uptake and effects of contaminants at some polluted sites.

OECD and the USEPA have used the zebra finch as an avian laboratory model for testing the toxicity of pesticides and other chemicals. The zebra finch has also been used as a surrogate for other songbirds for studying sublethal effects (e.g., immune response and sexual differentiation) of environmental contaminants. Research on the mechanisms of neurogenesis in the zebra finch will continue to be important for understanding neuronal processes in humans.

The mallard has been quite widely used for testing the effects of pesticides and other contaminants in the laboratory, and it has also been used to monitor the levels of contaminants (e.g., metals, PAHs) in the environment, as well as the effects of these contaminants on biochemical and physiological endpoints.

The American kestrel has been used as a laboratory model for testing the effects of pesticides and other environmental contaminants on the species and as a surrogate for other raptors. The American kestrel has also been used to monitor the levels and effects of environmental contaminants in the field.

The herring gull has been used for many studies on ecological health and ecotoxicology. The most extensive use of the herring gull for such studies has been in the Great Lakes of North America, where it continues to be the most studied avian species for monitoring the levels of environmental chemicals and their effects. In some studies, effects in gulls have been used to make predictions of possible effects in humans.

The European starling has been used in the field and in the laboratory for monitoring the effects of environmental contaminants. Starling nestlings are good biological monitors of local contamination, and monitoring of the levels of contaminants in this species and their effects can be used to evaluate the effects of remediation of contaminated sites. The development of better biochemical and molecular biomarkers in this species would be possible if more genetic sequence data were available.

2.5.6 NEMATODE MODELS

The nematode *Caenorhabditis elegans* is a small (~1 to 1.5 mm in length), free-living bactiverous soil nematode of the order Rhabditida. Into the mid to late 1960s, it was isolated and characterized by Sidney Brenner (1974). Since that time, all cell lineages of *C. elegans* have been traced (Sulston et al. 1983), its nervous system has been mapped (White et al. 1986), and it has served as the model organism for the Human Genome Project, resulting in its becoming the first multicellular organism to have a fully sequenced genome (*C. elegans* Sequencing Consortium 1998). These important accomplishments have combined to make it the most thoroughly studied organism currently known to science.

Analogous to the mammalian models, *C. elegans* meets the basic criteria to serve as a surrogate species. In core biological and human health studies, it has been used extensively to elucidate the underlying mechanisms of gene regulation, cell function, and developmental biology. *C. elegans* has been used as a model for ageing for more than 20 years (Zuckerman and Himmelhoch 1980). The transgenic capabilities of *C. elegans* have also made it useful as a model for human degenerative neurological diseases (Faber et al. 1999; Link 2001). It has also found use as a model in the biological response to anesthetic agents (Morgan and Cascorbi 1985; Crowder et al. 1996; Kayser et al. 1998).

C. elegans has been used to study additional human diseases including Menkes and Wilson's diseases, Alzheimer's disease, neurological disorders, polycystic kidney disease, Down syndrome, Huntington's disease, innate immunity, and cancer. Recently, the National Toxicology Program has developed *C. elegans* as an alternative test species to assess chemical toxicants. It is hoped that tests using *C. elegans* can contribute to models designed to understand human health risk assessment of individual environmental toxicants. Wild-type and transgenic strains of *C. elegans* have been used in environmental toxicity testing. Recently, the American Society for Testing and Materials (ASTM) adopted a standard guide of soil toxicity testing using this organism (ASTM 2001). Although there is limited comparative testing with other nematode species, studies have reported similar toxicities (Boyd and Williams 2003).

2.5.7 A COMMUNITY-BASED APPROACH FOR PROMOTING *DAPHNIA* AS A MODEL FOR ECOTOXICOGENOMICS

By many accounts, *Daphnia* is already regarded as an established model species for toxicological studies. Based on the wealth of accumulated data in ecotoxicological databases (Appendix A) and on its function as a test species for setting regulatory limits by environmental protection agencies (e.g., USEPA, Environment Canada, OECD), its relevance within the emerging field of toxicogenomics will grow in equal pace with the development of genomics resources. However, the costs and challenges associated with creating genomics tools for a new species are immense, and the promotion of this organism as a surrogate species requires the support of a large research community that can maximize the use of these resources.

Species of this freshwater crustacean, known as the water-flea, have fascinated biologists for the past 200 years within a variety of fields, including toxicology. These animals are common within all continents of the globe and are considered keystone species within aquatic habitats because of their central role within food-chains. Many of their biological attributes coincide with criteria for selecting surrogate species for ecological issues outlined in Table 2.2, counting well-defined population boundaries, ease in sampling large numbers at all life stages, and their sensitivity to chemical and ecological perturbations (Vaal et al. 1997a, 1997b). There are only a few reports of *Daphnia* being used for research to understand human health (e.g., Campbell et al. 2004). However, *Daphnia* are often used as indicator species in assessing the state of the environment (USEPA 1993; Stemberger and Lazorchek 1994). For example, specific assemblages of the zooplanktonic community typify water chemistry and can be used to gauge water quality and contaminant distribution within aquatic systems (Stemberger and Lazorchek 1994; Stemberger and Chen 1998; Chen et al. 2000). While these attributes add to their success as a model species for field studies, their biology is also ideal for experimental genetics in the lab.

Daphnia reproduce by cyclical parthenogenesis, which is characterized by their ability to reproduce clonally under favorable environmental conditions, thus creating genetically identical daughters. However, when cued by environmental signals linked to adverse conditions, mothers switch from producing daughters to producing sons and invest in haploid eggs requiring fertilization for development. This situation presents many unique opportunities needed to dissect genetic (among-clone) responses to ecological challenges from measurements that involve experimental and developmental noise (within-clone).

First, *Daphnia* are easily cultured and the clonal phase of their reproduction is rapid, taking 5 to 10 days for offspring to reach maturity. Moreover, brood sizes can be large. During recent years, experimentalists have used this system for measuring the rates of acclimation (in generational units) to various conditions and compounds (Muyssen and Janssen 2002, 2004; Bossuyt and Janssen 2004; Folt et al. 2004; Guan and Wang 2004) and for measuring the effects of acclimation and even accumulated mutations on fitness, similar to experiments on nematodes — for example, Vassilieva and Lynch (Vassilieva et al. 2000). Second, estimates of the genetic component of the variation in traits of interest are more accurately obtained using *Daphnia* because, after outbreeding the clonal lineages by inducing sexual reproduction in the laboratory, independent measurements are made on multiple individuals having the same genotype (Morgan et al. 2001). Without additional effort, the effects of inbreeding on traits and fitness can also be obtained by allowing matings between genetically identical males and females. These parameters are important if our goal is to understand the genetics for survival and evolutionary adaptation of natural populations exposed to toxicological stress, especially if demography is a key determinant (i.e., effective population size can dictate the amount of genetic variation).

2.5.7.1 The *Daphnia* Genomics Consortium (DGC)

The DGC was formed in October 2002 with the stated goal to "develop the *Daphnia* system to the same depth of molecular, cell and developmental biological understanding as other model systems, but with the added advantage of being able to interpret observations in the context of natural ecological challenges" (http://daphnia. cgb.indiana.edu). As of May 2005, the consortium consists of more than 80 members from 9 countries. Most of the early efforts focus on creating resources for *Daphnia pulex*. Although a distantly related species (*D. magna*) is more commonly used for toxicological research (Appendix A), *D. pulex* was first chosen because its natural history is pertinent to a greater number of investigators:

1) Its geographic range is vast compared to other narrowly endemic taxa in North America (Hebert 1995).
2) It is closely allied to a "complex" of hybridizing species that have adapted to live within a greater diversity of habitats within only the last few million years (Colbourne et al. 1998).

3) In certain lineages, the sexual phase of their reproduction is altogether lost, therefore enabling comparative studies on the consequences of shuffling the genome by recombination (Paland et al. 2005).
4) Within other lineages, polyploids of different ages can be studied for insights into the molecular evolution of duplicated genes and genomes.
5) They exhibit a greater range of phenotypic responses to environmental cues.
6) A number of studies also suggest that *D. pulex* is generally a more sensitive species to toxicants (Koivisto et al. 1992; Koivisto 1995; Shaw et al. 2005).

Nevertheless, the DGC is committed to bring *D. magna* on board as rapidly as possible. Although *D. magna* and *D. pulex* are classified within the same genus, they are as genetically distinct as mammals belonging to different orders (Colbourne and Hebert 1996). Genomics knowledge from both species or from the addition of *Ceriodaphnia* will ultimately reinforce the utility of daphnids as a surrogate species for comparative toxicogenomics, especially as a phylogenetic outgroup for comparisons to the insect genomes.

2.5.7.2 Community Resources

Whether the research goal of the investigator is to identify a sensitive and accurate genomics signature of specific cellular stress or to understand the adaptive functions of an organism as an account of how genes, traits, and the environment interact to affect survival and reproduction, the task begins by identifying genes that encode a trait of interest. Three DNA-based approaches can accomplish this first mission. Each approach requires a specific set of genomics and bioinformatic tools, which the DGC is providing.

First, a candidate gene approach may be useful and readily applied if the genes affecting the trait of interest are already known and conserved in other model species. For the *Daphnia* researcher, this method is practical because of the well characterized and fully sequenced genome of the fellow arthropod *Drosophila*. Arguably, the fruit fly is the best understood eukaryote in its assignment of function to genes. Of the 13,472 annotated protein coding genes (Release 4.0), 60% have been ascribed a molecular function and a biological process, although this figure includes nontraceable reports. Therefore, a minimal community toolkit consisting of a searchable genome sequence database and of arrayed genomics DNA libraries creates important research opportunities. The DGC has a *Daphnia* genome database called wFleaBase, which is designed to curate and share rapidly accumulating genetic, molecular, and functional genomics data (Colbourne et al. 2005).

Second, gene position mapping by quantitative-trait locus (QTL) analysis can potentially focus the position of a wanted gene within the genome (Lynch and Walsh 1998). For the *Daphnia* system, the DGC has developed a battery of codominant polymorphic markers called microsatellites that are currently being genetically and physically mapped to chromosomes (Colbourne et al. 2004). When the maps are complete, a relationship between the genetic linkage among markers and their physical distance on the chromosomes will be drawn to produce a useful set of markers for distribution to the community that will minimally bracket the position of genes affecting specific traits.

A principal advantage of locating genes by QTL analysis is shared with a third gene discovery method using microarrays. Both approaches can identify genes with no known functions, which is important given that even for the best model eukaryotes, roughly 40–50% of the predicted genes have yet to be characterized. In this regard, the DGC is creating cDNA microarrays for *D. pulex* and *D. magna* (http://daphnia.cgb.indiana.edu/). To ensure that a diversity of genes is available for probes on the arrays, cDNA libraries enriched for full-length transcripts are created from a clonal assemblage of animals exposed to 10 separate ecological challenges, including environmental toxins (Bauer et al. 2004).

Additional tools are being developed that will likely provide opportunities to validate results by genetic experimentation. For example, DGC members are busy producing cell lines for basic biological studies and transformation (reverse genetics) in *Daphnia*. Others are producing bioluminescent and fluorescent genes for in situ hybridization experiments (http://daphnia.cgb.indiana.edu/projects). However, the ultimate resource for an enterprise to discover toxicologically relevant genes is an annotated genome sequence for *Daphnia*.

2.5.7.3 The *Daphnia* Genome Project

The *Daphnia* project is a multi-institutional endeavor. Members of the DGC identified the isolate to be sequenced from a survey of populations inhabiting temporary ponds in the state of Oregon during the course of a population genetic study (Lynch and Walsh 1998). From this survey, a population was found to contain very little genetic variation (~4% heterozygosity) and additional sequencing discovered a single naturally occurring clonal isolate possessing the lowest nucleotide heterozygosity (~0.14%). This level of variation is sufficiently small to permit a successful assembly of the genome derived by shotgun sequencing at the Joint Genome Institute (U.S. Department of Energy). Yet, to facilitate the assembly and annotation, the sequencing is supported by 5 other projects at various DGC member institutions (Figure 2.4): 1) *Daphnia* DNA cloned within a bacterial artificial chromosome (BAC) library with 10× coverage of the genome (insert sizes of ~150 Kb) is being tiled into chromosomal stretches of DNA using restriction enzymes. By matching the restriction site patterns shared among BACs, they are aligned to one another in sequential order without sequencing. Then, by matching the same restriction site patterns shared among contiguous BACs with the genomics sequence, gaps among orphan sequences are identified. 2) The ordered markers on the genetic map are matched to the sequence scaffolds to also orient orphan sequences in the genome assembly. 3) The cDNA sequencing project, aiming to characterize a diversity of genes for a *Daphnia* gene collection and 4) microarrays, will help annotate the genome. 5) In the end, all information for *Daphnia* obtained from the assembly of its genome, the annotation of its genes, and the discovery of gene function by microarray experiments is deposited in the *Daphnia* genome database.

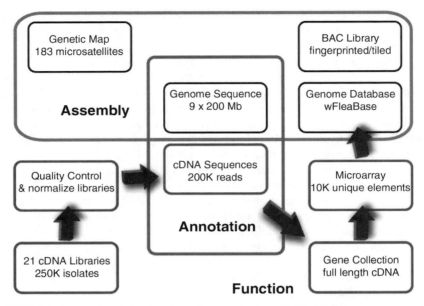

FIGURE 2.4 Schematic of the *Daphnia* Genome Project highlighting how a community-based approach accelerates the creation of necessary resources for toxicogenomics research. Five projects (i–v) at various *Daphnia* Genomics Consortium (DGC) member institutions support the whole-genome sequencing project of *D. pulex* at the Joint Genome Institute for assembling, annotating, and ascribing function based on condition-dependent expression patterns. Small boxes represent work conducted by various project leaders (http://daphnia. cgb.indiana.edu). Data from (i) fingerprinting and tiling a BAC library; and (ii) genetic mapping and genome shotgun sequencing are assembled to ultimately create genomics scaffolds representing whole chromosomes, which are made accessible to the research community in genome databases. Data from (iii) cDNA sequencing serve to annotate the genome assembly, to create a *Daphnia* Gene Collection as full-length transcripts within plasmid libraries, and to provide probes printed on (iv) microarrays. These resources in turn are used to elucidate the function of genes. All data are deposited in the (v) *Daphnia* genome database.

2.6 CONCLUSIONS

For field-based comparative ecotoxicogenomics, toxicologists might focus on non-mammalian (e.g., aquatic and avian) models, and then compare these data to mammalian laboratory data or human studies. Various economic and ethical factors support reducing the number of mammals used in biomedical research, making the use of nonmammalian models attractive. As more nonmammalian models are characterized and genomes sequenced, the power of the comparative approach increases (Ballatori and Villalobos 2002). Recent genome sequencing projects have demonstrated that the human genome is remarkably similar to evolutionarily distant organisms. The similarities between invertebrate and vertebrate species, including humans, support the notion that studies in an assortment of organisms can provide key insights into health and ecological science and risk assessment. The fact that many evolutionarily divergent organisms have relatively smaller genomes is yet another compelling reason for using such models in comparative toxicogenomics, and the growing portfolio of genomes makes this synthesis possible.

The Comparative Toxicogenomics Database (CTD) is one promising tool being developed to facilitate comparative studies of toxicologically significant genes and proteins across diverse species (Mattingly et al. 2004; http://ctd.mdibl.org/). Of the tens of thousands of chemicals currently used and regulated across the planet, the toxic potential and the molecular mechanisms underlying the action of most of these chemicals are not well understood. With the rising number of publicly available sequences and completely sequenced genomes, comparative toxicogenomics studies could be a key for elucidating biological systems and annotating genomics and proteomics data. Comparisons of distantly related vertebrate and invertebrate species may be of particular value for identifying conserved genetic and molecular mechanisms significant to various toxicities. The CTD is being developed with an emphasis on aquatic and mammalian species; the goals are to provide unique insights into conserved sequences and polymorphisms and to inform the integration of health and ecological risk assessments.

In this chapter, we have identified and prioritized a suite of fewer than 20 species that can serve as surrogates for environmental and human health and that could serve as an initial starting point for a multinational comparative toxicogenomics research program. In accomplishing this goal, a set of criteria for selecting surrogate species was defined that is applicable to future iterations of the process. These criteria can be applied in reviewing species currently used in toxicological testing, ecological monitoring and research, and the genomics sciences, evaluating whether these species represent a model for diseases and outcomes outside their class. Genomics greatly enables this comparative approach, affording quantification of common response and discovery of toxicity pathways shared among species. The toxicological test species identified here would bridge the gaps between current test and monitoring species; species critical to understanding and appreciating the effects of chemical and environmental stressors; and species for which genomes have been sequenced.

REFERENCES

Amaya E, Kroll K. 1999. A method for generating transgenic frog embryos. Methods Molec Biol 97:393–414.

AmphibiaWeb. www.amphibiaweb.org/aw/amphibian/speciesnums.html. Accessed 18 May 2005.

Amsterdam A, Nissen RM, Sun Z, Swindell EC, Farrington S, Hopkins N. 2004 Identification of 315 genes essential for early zebrafish development. Proc Natl Acad Sci U S A 101:12792–12797.

Anderson GL, Cole RD, Williams PL. 2004. Assessing behavioral toxicity with *Caenorhabditis elegans*. Environ Toxicol Chem 23:1235–1240.

Aparicio S, Chapman J, Stupka E, Putnam N, Chia J-M, Dehal P, Christoffels A, Rash S, Hoon S, Smit A, and others. 2002. Whole-genome shotgun assembly and analysis of the genome of *Fugu rubripes*. Science 297:1301–1310.

Armitage D. 2004 *Rattus norvegicus.* Animal Diversity Web. http://animaldiversity.ummz. umich.edu/site/accounts/information/Rattus_norvegicus.html. Accessed 04 Nov 2004.

[ASTM] American Society for Testing and Materials. 2002. Standard guide for conducting laboratory soil toxicity tests with the nematode *Caenorhabditis elegans.* E 2172-01. Annual Book of ASTM Standards, Vol 11.05. West Conshohocken, PA. p 1606–1616.

Azumi K, Fujie M, Usami T, Miki Y, Satoh N. 2004. A cDNA microarray technique applied for analysis of global gene expression profiles in tributyltin-exposed ascidians. Marine Environ Res 58:543–546.

Ballatori N, Villalobos AR. 2002. Defining the molecular and cellular basis of toxicity using comparative models. Toxicol Applied Pharmacol 183:207–220.

Ballenger L. 1999. *Mus musculus,* Animal Diversity Web. http://animaldiversity.ummz.umich. edu/site/accounts/information/Mus_musculus.html. Accessed 04 November 2004.

Barr MM, Sternberg PW. 1999. A polycystic kidney-disease gene homologue required for male mating behaviour in *C. elegans.* Nature 401:386–389.

Bauer D, Bohuski E, Colbourne JK, Haney JF, Lynch M, Thomas WK. 2004 The *Daphnia* Genomics Consortium: development of the *Daphnia* gene collection. Genes in Ecology, Ecology in Genes Conference, Overland Park, KS.

Berghmans S, Jette C, Langenau D, Hsu K, Stewart R, Look T, Kanki JP. 2005. Making waves in cancer research: new models in the zebrafish. Biotechniques 39:227–237.

Boyd WA, Williams PL. 2003. Availability of metals to the nematode *Caenorhabditis elegans:* toxicity based on total concentrations in soil and extracted fractions. Environ Toxicol Chem 22:1100–1106.

Brenner S. 1974. The genetics of *Caenorhabditis elegans.* Genetics 77:71–94.

Brenner S. 1988. Foreword. In: Wood WB, editor. The nematode *Caenorhabditis elegans.* Plainview, NY: Cold Spring Harbor Laboratory. p ix–xiii.

Bossuyt BT, Janssen CR. 2004. Influence of multigeneration acclimation to copper on tolerance, energy reserves, and homeostasis of *Daphnia magna* straus. Environ Toxicol Chem 23:2029–2037.

Cai SY, Wang L, Ballatori N, Boyer JL. 2001. Bile salt export pump is highly conserved during vertebrate evolution and its expression is inhibited by PFIC type II mutations. Am J Physiol Gastrointest Liver Physiol 281:G316–G322.

C. elegans Sequencing Consortium. 1998. Genome sequence of the nematode *C. elegans*: a platform for investigating biology. Science 282:2012–2018.

Campbell AK, Wann KT, Matthews SB. 2004. Lactose causes heart arrhythmia in the water flea *Daphnia pulex.* Comp Biochem Physiol B Biochem Mol Biol 139:225–234.

Chen CY, Stemberger RS, Klaue B, Blum JD, Pickhardt PC, Folt CL. 2000. Accumulation of heavy metals in food web components across a gradient of lakes. *Limnol Oceanog* 45:1525–1536.

Colbourne JK, Crease TJ, Weider LJ, Hebert PDN, Dufresne F, Hobaek A. 1998. Phylogenetics and evolution of a circumarctic species complex (Cladocera: *Daphnia pulex*). Biol J Linnean Soc 65:347–365.

Colbourne JK, Hebert PD. 1996. The systematics of North American *Daphnia* (Crustacea: Anomopoda): a molecular phylogenetic approach. Philos Trans R Soc Lond B Biol Sci 351:349–360.

Colbourne JK, Robison B, Bogart K, Lynch M. 2004. Five hundred and twenty-eight microsatellite markers for ecological genomic investigations using *Daphnia.* Molec Ecol Notes 4:485–490.

Colbourne JK, Singan VR, Gilbert DG. 2005. wFleaBase: the *Daphnia* genome database. BMC Bioinformatics 6:45.

Cole RD, Anderson GL, Williams PL. 2004. The nematode *Caenorhabditis elegans* as a model of organophosphate-induced mammalian neurotoxicity. Toxicol Appl Pharmacol 194:248–256.

Courchesne C, Bantle J. 1985. Analysis of the activity of DNA, RNA, and protein synthesis inhibitors on Xenopus embryo development. Teratog Carcinog Mutagen 5:177–193.

Crowder CM, Shebester LD, Schedl T. 1996. Behavioral effects of volatile anesthetics in *Caenorhabditis elegans*. Anesthesiology 85:901–912.

Crump D, Werry K, Veldhoen N, Van Aggelen G, Helbing CC. 2002. Exposure to the herbicide acetochlor results in increased thyroid hormone-dependent gene expression and accelerated metamorphosis in *Xenopus laevis*. Environ Health Perspect 110:1199–1205.

Dehal P, Dehal P, Satou Y, Campbell RK, Chapman J, et al. 2002. The draft genome of *Ciona intestinalis:* Insights into chordate and vertebrate origins. Science 298:2157–2167.

Drummond IA. 2005. Kidney development and disease in the zebrafish. J Am Soc Nephrol 16:299–304.

Epstein FH, Epstein JA. 2005. A perspective on the value of aquatic models in biomedical research. Exp Biol Med (Maywood) 230:1–7.

Faber PW, Alter JR, MacDonald ME, Hart AC. 1999. Polyglutamine-mediated dysfunction and apoptotic death of a *Caenorhabditis elegans* sensory neuron. Proc Natl Acad Sci USA 96:179–184.

Folt CL, Glaholt SP, Chen CY, Hamilton JW, Shaw JR. 2004. Metal tolerance in *Daphnia pulex*. Proceedings from 25th Annual Meeting of the Society of Environmental Toxicology and Chemistry. Portland, OR: November 14–18.

Forrest JN Jr. 1996. Cellular and molecular biology of chloride secretion in the shark rectal glands: regulation by adenosine receptors. Kidney Int 49:1557–1562.

Genome Sequencing Project Consortium. 2004. Genome sequence of the Brown Norway rat yields insights into mammalian evolution. Nature 428:493–521.

Gracey AY, Fraser EJ, Li W, Fang Y, Taylor RR, Rogers J, Brass A., Cossins AR. 2004. Coping with cold: An integrative, multitissue analysis of the transcriptome of a poikilothermic vertebrate. Proc Natl Acad Sci U S A 101:16970–16975.

Gray LE Jr, Wilson V, Noriega N, Lambright C, Furr J, Stoker TE, Laws SC, Goldman J, Cooper RL, Foster PM. 2004. Use of the laboratory rat as a model in endocrine disruptor screening and testing. ILAR J 45:425–437.

Gregory TR. 2005. Animal Genome Size Database. http://www.genomesize.com

Guan R, Wang WX. 2004. Cd and Zn uptake kinetics in *Daphnia magna* in relation to Cd exposure history. Environ Sci Technol 38:6051–6058.

Hamadeh HK, Bushel PR, Jayadev S, Martin K, DiSorbo O, Sieber S, and others. 2002. Gene expression analysis reveals chemical-specific profiles. Toxicol Sci 67:219–231.

Hebert PDN. 1995 The *Daphnia* of North America. CD-ROM distributed by author. Department of Zoology, University of Guelph.

Helbing CC, Werry K, Crump D, Domanski D, Veldhoen N, Bailey CM. 2003. Expression profiles of novel thyroid hormone-responsive genes and proteins in the tail of *Xenopus laevis* tadpoles undergoing precocious metamorphosis. Molec Endocrinol 17:1395–1409.

Hill AJ, Teraoka H, Heideman W, Peterson RE. 2005. Zebrafish as a model vertebrate for investigating chemical toxicity. Toxicol Sci 86:6–19.

International Mouse Strain Resource. 2004. http://www.informatics.jax.org/imsr/index.jsp, Accessed 4 November 2004.

Jahnke GD, Choksi NY, Moore JA, Shelby MD. 2004. Thyroid toxicants: assessing reproductive health effects. Environ Health Perspect 112:363–368.

Katsiadaki I, Scott AP, Hurst MR, Matthiessen P, Mayer I. 2002. Detection of environmental androgens: a novel method based on enzyme-linked immunosorbent assay of spiggin, the stickleback (*Gasterosteus aculeatus*) glue protein. Environ Toxicol Chem 21:1946–1954.

Katagiri T, Asakawa S, Minagawa S, Shimizu N, Hirono I, Aoki T. 2001. Construction and characterization of BAC libraries for three species; rainbow trout, carp and tilapia. Int Soc Animal Genet Animal Genet 32:200–204.

Kayser B, Rajaram S, Thomas S, Morgan PG, Sedensky MM. 1998. Control of anesthetic response in C. elegans. Toxicol Lett 100-101:339–346.

Keller ET, Murtha JM. 2004. The use of mature zebrafish (Danio rerio) as a model for human aging and disease. Comp Biochem Physiol C Toxicol Pharmacol 138:335–341.

Koivisto S. 1995. Is *Daphnia-magna* an ecologically representative zooplankton species in toxicity tests? Environ Pollut 90:263–267.

Koivisto S, Ketola M, Walls M. 1992. Comparison of 5 cladoceran species in short-term and long-term copper exposure. Hydrobiologia 248:125–136.

Lambrechts D, Carmeliet P. 2004. Genetics in zebrafish, mice, and humans to dissect congenital heart disease: insights in the role of VEGF. Curr Top Dev Biol 62:189–224.

Larkin P, Folmar LC, Hemmer MJ, Poston AJ, Denslow ND. 2003. Expression profiling of estrogenic compounds using a sheepshead minnow cDNA macroarray. Environ Health Perspect Toxicogenom 111:839–846.

Link CD, Taft A, Kapulkin V, Duke K, Kim S, Fei Q, Wood DE, Sahagan BG. 2003. Gene expression analysis in a transgenic *Caenorhabditis elegans* Alzheimer's disease model. Neurobiol Aging 24:397–413.

Lynch M, Walsh JB. 1998. Genetics and analysis of quatitative traits. Sunderland, MA: Sinauer Associates.

Mattingly CJ, Colby GT, Rosenstein MC, Forrest Jr JN, Boyer JL. 2004. Promoting comparative molecular studies in environmental health research: an overview of the comparative toxicogenomics database (CTD). Pharmacogenomics J4:5–8.

McMahon C, Semina EV, Link BA. 2004. Using zebrafish to study the complex genetics of glaucoma. Comp Biochem Physiol C Toxicol Pharmacol 138:343–350.

Morgan KK, Hicks J, Spitze K, Latta L, Pfrender ME, Weaver CS, Ottone M, Lynch M. 2001. Patterns of genetic architecture for life-history traits and molecular markers in a subdivided species. Evolution 55:1753–1761.

Morgan PG, Cascorbi HF. 1985. Effect of anesthetics and a convulsant on normal and mutant *Caenorhabditis elegans*. Anesthesiology 62:738–744.

Muyssen BT, Janssen CR. 2002. Multi-generation cadmium acclimation and tolerance in *Daphnia magna* Straus. Environ Pollut 130:309–316.

Muyssen BT, Janssen CR. 2002. Accumulation and regulation of zinc in *Daphnia magna*: links with homeostasis and toxicity. Arch Environ Contam Toxicol 43:492–496.

Nairn RS, Kazianis S, Della Coletta L, Trono D, Butler AP, Walter RB, Morizot DC. 2001. Genetic analysis of susceptibility to spontaneous and UV-induced carcinogenesis in Xiphophorus hybrid fish. Mar Biotechnol 3(Suppl 1):S24–S36.

[NCBI] National Center for Biotechnology Information. 2004. Rat Genome Resources. http://www.ncbi.nlm.nih.gov/genome/guide/rat/index.html, Accessed 04 November 2004.

Oleksiak ME, Churchill GA, Crawford DL. 2002. Variation in gene expression within and among natural populations. *Nature Genetics* 32:261–266.

Paland S, Colbourne JK, Lynch M. 2005. Evolutionary history of contagious asexuality in *Daphnia pulex*. Evolution 59:800–813.

Palti Y, Gahr SA, Hansen JD, Rexroad III CE. 2004. Characterization of a new BAC library for rainbow trout: evidence for multi-locus duplication. Int Soc Animal Genet Animal Genet 35:130–133.

Pennisi E. 2003. Modernizing the tree of life. Science 300:1692–1697.

Pough F, Andrews R, and others. 2001. Herpetology. Upper Saddle River, NJ: Prentice-Hall.

Rat Genome Database. 2004. http://rgd.mcw.edu/. Accessed 4 November 2004.

Rise ML, von Schalburg KR, Brown GD, Mawer MA, Devlin RH, Kuipers N, Busby M, Beetz-Sargent M, Alberto R, Gibbs AR, and others. 2004. Development and application of a salmonid EST database and cDNA microarray: data mining and interspecific hybridization characteristics. Genome Res 14:478–490.

Sakamoto D, Kudo H, Inohaya K, Yokoi H, Narita T, Naruse K, Mitani H, Araki K, Shima A, Ishikawa Y, et al. 2004. A mutation in the gene for delta-aminolevulinic acid dehydratase (ALAD) causes hypochromic anemia in the medaka, *Oryzias latipes*. Mech Dev 121:747–752.

Schluter D. 2004 Parallel evolution and inheritance of quantitative traits. Am Nat 63:809–822.

Schwartz DA, Freedman JH, Linney EA. 2004. Environmental genomics: a key to understanding biology, pathophysiology and disease. Hum Mol Genet 13(Spec No 2):R217–R224.

Shaw JR, Dempsey TD, Chen CY, Hamilton JW, Folt CL. 2005. The comparative toxicity of cadmium, zinc and mixtures of cadmium and zinc to Daphniids. Environ Toxicol Chem. 25:182–189.

Sheader DL, Gensberg K, Lyons BP, Chipman K. 2004. Isolation of differentially expressed genes from contaminant exposed European flounder by suppressive, subtractive hybridisation. Mar Environ Res 58:553–557.

Stanford Genome Evolution Center. http://cegs.stanford.edu/Stickleback_course_poster.jsp, Accessed 27 May 2005.

Stemberger RS, Chen CY. 1998. Fish tissue metals and zooplankton assemblages of northeastern US lakes. Can J Fish Aquat Sc. 55:339–352.

Stemberger RS, Lazorchek JM. 1994. Zooplankton assemblage responses to disturbance gradients. Can J Fish Aquat Sci 51:2435–2447.

Sulston JE, Schierenberg E, White JG, Thompson JN. 1983. The embryonic cell lineage of the nematode *Caenorhabditis elegans*. Dev Biol 100:64–119.

Taylor MR, Hurley JB, Van Epps HA, Brockerhoff SE. 2004. A zebrafish model for pyruvate dehydrogenase deficiency: rescue of neurological dysfunction and embryonic lethality using a ketogenic diet. Proc Natl Acad Sci USA 101:4584–4589.

Thorgaard GH, Bailey GS, Williams D, Buhler DR, Kaattari SL, Ristow SS, Hansen JD, Winton JR, Bartholomew JL, Nagler JJ, et al. 2002. Status and opportunities for genomics research with rainbow trout. Comp Biochem and Physiol Part B 133:609–646.

[USEPA] US Environmental Protection Agency. 1993. Methods for measuring the acute toxicity of effluents and receiving waters to freshwater and marine organisms. EPA/600/4-89/001.

[USEPA] US Environmental Protection Agency. 2004. Harmonized test guidelines, Office of Prevention Pesticides Toxic Substances. http://www.epa.gov/opptsfrs/home/guidelin.htm. Accessed 4 November 2004.

Vaal M, van der Wal JT, Hermens J, Hoekstra J. 1997a. Pattern analysis of the variation in the sensitivity of aquatic species to toxicants. Chemosphere 35:1291–1309.

Vassilieva LL, Hook AM, Lynch M. 2000. The fitness effects of spontaneous mutations in *Caenorhabditis elegans*. Evolution 54:1234–1246.

Veldhoen N, Skirrow RC, Ji L, Domanski D, Bonfield ER, Bailey CM, Helbing CC. 2006. Use of heterologous cDNA arrays and organ culture in the detection of thyroid hormone-dependent responses in a sentinel frog, *Rana catesbeiana*. Comp Biochem Physiol. Part D, 1:187–199.

Web of Science. 2005. Search terms were "Rana" OR "Bufo". Accessed 18 May 2005.

White JG, Southgate E, Thompson JN, Brenner S. 1986. The structure of the nervous system of the nematode *Caenorhabditis elegans.* Philos Trans R Soc London B 314:1–340.

Zuckerman BM, Himmelhoch S. 1980. Nematodes as models to study aging. In: Zuckerman BM, editor. Nematodes as biological models. Vol. 2. Aging and other models. New York: Academic Press. p. 4–28.

APPENDIX A

Rankings of top 25 aquatic animals[a] and top 25 terrestrial animals[b] according to number of records in ECOTOX database[c]

Rank	Species	No. records	Percent records	Cumulative percentage
	Aquatic animals			
1	*Oncorhynchus mykiss*	16,261	10.5	10.5
2	*Daphnia magna*	10,096	6.5	17.0
3	*Pimephales promelas*	8885	5.7	22.8
4	*Lepomis macrochirus*	6691	4.3	27.1
5	*Cyprinus carpio*	3736	2.4	29.5
6	*Oryzias latipes*	2960	1.9	31.4
7	*Oncorhynchus kisutch*	2586	1.7	33.1
8	*Ictalurus punctatus*	2506	1.6	34.7
9	*Carassius auratus*	2338	1.5	36.3
10	*Poecilia reticulata*	2216	1.4	37.7
11	*Petromyzon marinus*	1867	1.2	38.9
12	*Cyprinodon variegatus*	1865	1.2	40.1
13	*Oncorhynchus tshawytscha*	1841	1.2	41.3
14	*Danio rerio*	1739	1.1	42.4
15	*Ceriodaphnia dubia*	1592	1.0	43.4
16	*Ptychocheilus oregonensis*	1511	1.0	44.4
17	*Mytilus edulis*	1285	0.8	45.2
18	*Salvelinus fontinalis*	1241	0.8	46.1
19	*Tilapia mossambica*	1189	0.8	46.8
20	*Daphnia pulex*	1134	0.7	47.6
21	*Salmo salar*	1072	0.7	48.2
22	*Americamysis bahia*	1044	0.7	48.9
23	*Gambusia affinis*	979	0.6	49.6
24	*Xenopus laevis*	945	0.6	50.2
25	*Crassostrea virginica*	890	0.6	50.7
	Terrestrial animals			
1	*Anas platyrhynchos*	1344	13.9	13.9
2	*Colinus virginianus*	1319	13.6	27.4
3	*Eisenia fetida*	653	6.7	34.2
4	*Rattus norvegicus*	637	6.6	40.7
5	*Apis mellifera*	535	5.5	46.3
6	*Mus musculus*	510	5.3	51.5
7	*Coturnix coturnix japonica*	437	4.5	56.0

Rank	Species	No. records	Percent records	Cumulative percentage
8	*Phasianus colchicus*	275	2.8	58.9
9	*Gallus domesticus*	265	2.7	61.6
10	*Folsomia candida*	233	2.4	64.0
11	*Peromyscus maniculatus*	171	1.8	65.7
12	*Nemata*	140	1.4	67.2
13	*Gryllus pennsylvanicus*	140	1.4	68.6
14	*Lumbricus terrestris*	125	1.3	69.9
15	*Lumbricus rubellus*	114	1.2	71.1
16	*Eisenia andrei*	98	1.0	72.1
17	*Lumbricidae*	78	0.8	72.9
18	*Caenorhabditis elegans*	67	0.7	73.6
19	*Mus* spp.	62	0.6	74.2
20	*Passer domesticus*	54	0.6	74.8
21	*Agelaius phoeniceus*	50	0.5	75.3
22	*Blattella germanica*	39	0.4	75.7
23	*Columba livia*	38	0.4	76.1
24	*Musca domestica*	37	0.4	76.5
25	*Mustela vison*	34	0.4	76.8

[a] From a total of 4205 species.

[b] From a total of 1817 species.

[c] Relative and cumulative contributions to toxicological information are denoted as a percentage of respective databases.

APPENDIX B

Animal genome sequencing projects as of August 2004

Species	Group	Status
Aedes aegypti	Arthropoda	In progress
Aedes albopictus	Arthropoda	In progress
Aedes triseriatus	Arthropoda	In progress
Anopheles gambiae str. PEST	Arthropoda	Complete
Apis mellifera	Arthropoda	In assembly
Biomphalaria glabrata	Mollusks	In progress
Bombyx mori	Arthropoda	Complete
Bos taurus	Mammals	In progress
Branchiostoma floridae	Cephalochordates	In progress
Brugia malayi	Nematodes	In progress
Caenorhabditis briggsae	Nematodes	Complete
Caenorhabditis elegans	Nematodes	Complete
Caenorhabditis japonica	Nematodes	In progress
Caenorhabditis remanei	Nematodes	In progress
Caenorhabditis sp. CB5161	Nematodes	In progress
Canis familiaris	Mammals	In assembly

Species	Group	Status
Canis latrans	Mammals	In progress
Canis lupus	Mammals	In progress
Cavia porcellus	Mammals	In progress
Ciona intestinalis	Urochordates	Complete
Ciona savignyi	Ascidian	In assembly
Culex pipiens	Arthropoda	In progress
Danio rerio	Fish	In progress
Daphnia pulex	Crustaceans	In progress
Dasypus novemcinctus	Mammals	In progress
Drosophila ananassae	Arthropoda	In progress
Drosophila erecta	Arthropoda	In progress
Drosophila grimshawi	Arthropoda	In progress
Drosophila melanogaster	Arthropoda	Complete
Drosophila mojavensis	Arthropoda	In progress
Drosophila persimilis	Arthropoda	In progress
Drosophila pseudoobscura	Arthropoda	In assembly
Drosophila sechellia	Arthropoda	In progress
Drosophila simulans	Arthropoda	In progress
Drosophila virilis	Arthropoda	In progress
Drosophila willistoni	Arthropoda	In progress
Drosophila yakuba	Arthropoda	In assembly
Echinops telfairi	Mammals	In progress
Equus caballus	Mammals	In progress
Erinaceus europaeus	Mammals	In progress
Felis catus	Mammals	In progress
Gallus gallus	Birds	In assembly
Gasterosteus aculeatus	Fish	In progress
Glossina morsitans	Arthropoda	In progress
Heliothis virescens	Arthropoda	In assembly
Homo sapiens	Primates	Complete
Hydra magnipapillata	Cnidaria	In progress
Lemur catta	Primates	In progress
Loxodonta africana	Mammals	In progress
Lytechinus variegatus	Echinozoa	In progress
Macaca mulatta	Primates	In progress
Macropus eugenii	Mammals	In progress
Meleagris gallopavo	Birds	In progress
Meloidogyne hapla	Nematodes	In progress
Monodelphis domestica	Mammals	In progress
Mus musculus	Mammals	Complete
Nematostella vectensis	Cnidaria	In progress
Oikopleura dioica	Appendicularians	In progress
Oreochromis niloticus	Fish	In progress
Ornithorhynchus anatinus	Mammals	In progress
Oryctolagus cuniculus	Mammals	In progress
Ovis aries	Mammals	In progress
Pan troglodytes	Primates	In assembly
Papio anubis	Primates	In progress

Species	Group	Status
Petromyzon marinus	Fish	In progress
Pongo pygmaeus	Primates	In progress
Pristionchus pacificus	Nematodes	In progress
Rattus norvegicus	Mammals	Complete
Saccoglossus kowalesvskii	Hemichordates	In progress
Salmo salar	Fish	In progress
Schistosoma mansoni	Platyhelminthes	In progress
Sorex araneus	Mammals	In progress
Strongylocentrotus purpuratus	Echinozoa	In progress
Sus scrofa	Mammals	In progress
Takifugu rubripes	Fish	Complete
Tetraodon nigroviridis	Fish	In assembly
Tribolium castaneum	Arthropoda	In progress
Trichinella spiralis	Nematodes	In progress
Trichoplax adhaerens	Placozoa	In progress
Xenopus laevis	Amphibia	In progress
Xenopus tropicalis	Amphibia	In progress

APPENDIX C

Species relevant to human and environmental toxicology considered in this chapter

Group	Scientific name	Common name	Habitat
Amphibian	*Xenopus tropicalis*	West African clawed frog	fw
Amphibian	*Xenopus laevis*	South African clawed frog	fw
Amphibian	*Rana rugosa*	Wrinkled frog	fw/t
Amphibian	*Rana pipiens*	Northern leopard frog	fw/t
Amphibian	*Rana temporaria*	European common frog	fw/t
Amphibian	*Rana esculenta*	Edible frog	fw/t
Amphibian	*Bufo bufo*	European toad	fw/t
Amphibian	*Rana catesbeiana*	North American bullfrog	fw/t
Amphibian	*Ambystoma tigrinum*	Tiger salamander	fw
Annelid	*Lumbricus terrestris*	Common earthworm	s
Annelid	*Eisenia fetida*	Common brandling worm	s
Bird	*Taeniopygia guttata*	Zebrafinch	t
Bird	*Sterna hirundo*	Tern	t
Bird	*Sturnus vulgaris*	Starling	t
Bird	*Agelaius phoeniceus*	Red-wing blackbird	t
Bird	*Anas platyrhynchos*	Mallard duck	t
Bird	*Milvus migrans*	Kite	t
Bird	*Coturnix japonicus*	Japanese quail	t
Bird	*Passer domesticus*	House sparrow	t
Bird	*Larus argetatus*	Herring gull	t

Group	Scientific name	Common name	Habitat
Bird	*Taeniopygia guttata*	Zebra finch	t
Bird	*Phalacrocorax auritus*	Double-crested cormorant	m
Bird	*Gallus gallus*	Chicken	t
Bird	*Colinus virginianus*	Bobwhite quail	t
Bird	*Falco sparverius*	American kestrel	t
Bivalve	*Mytilus edulis*	Mussel	m
Chordate	*Ciona intestinales*	Sea squirt	m
Crustacean	*Brachionus calyciflorus*	Rotifer	fw
Crustacean	*Americamysis bahia*	Mysid shrimp	m
Crustacean	*Hyalella azteca*	Amphipod	fws
Crustacean	*Daphnia pulex*	Water flea	fw
Crustacean	*Daphnia magna*	Water flea	fw
Crustacean	*Ceriodaphnia dubia*	Water flea	fw
Crustacean	*Tigriopus* sp.	Copepod	m
Fish	*Danio rerio*	Zebrafish	fw
Fish	*Xiphophorus*	Swordtail	fw
Fish	*Tilapia*	Tilapia	fw
Fish	*Gasterosteus aculeatus*	Three-spined stickleback	fw
Fish	*Cyprinodon variegatus*	Sheepshead minnow	m
Fish	*Onchorhynchus mykiss*	Rainbow trout	fw
Fish	*Oryzias latipes*	Medaka	fw
Fish	*Micropterus salmoides*	Largemouth bass	fw
Fish	*Poecilia reticulata*	Guppy	fw
Fish	*Carassius auratus*	Goldfish	fw
Fish	*Fundulus heteroclitus*	Fundulus	m
Fish	*Paralichthys olivaceus*	Flounder	m
Fish	*Pimephales promelas*	Fathead minnow	fw
Fish	*Ictalurus punctatus*	Channel catfish	fw
Fish	*Cyprinus carpio*	Carp	fw
Fish	*Lepomis macrochirus*	Bluegill sunfish	fw
Fish	*Salmo salar*	Atlantic salmon	fw
Insect	*Folsomia candida*	Springtail	s
Insect	*Apis mellifera*	Honeybee	t
Insect	*Drosophila melanogaster*	Fruit fly	t
Mammals	*Rattus norvegicus*	Norway rat	t
Mammals	*Nyctereutes procyonoides*	Racoon dog	t
Mammals	*Oryctolagus cuniculus*	New Zealand white rabbit	t
Mammals		Nonhuman primates	t
Mammals	*Mus musculus*	House mouse	t
Mammals	*Homo sapiens*	Human	t
Mammals	*Canis familiaris*	Dog	t
Mammals	*Felis domesticus*	Cat	t
Nematode	*Caenorhabditis elegans*	Nematode worm	s
Other invertebrate	*Strongylocentrotus purpuratus*	Sea urchin	m
Other invertebrate	*Dictyostelium*	Slime mold	t
Other invertebrate	Cnidarians		fw, m
Insect	*Chironomus riparius*	Midge	fws/t

Group	Scientific Name	Common Name	Habitat
Reptile	*Pseudemys rubriventris*	Red-bellied turtle	t
Reptile	*Alligator mississipiensis*	Alligator	t
Yeast	*Saccharomyces cerevisae*	Yeast	t, fw, s, fws

Notes: t = terrestrial; fw = freshwater; fws = freshwater sediment; m = marine; ms = marine sediment; s = soil; fw/t = part of life cycle as an independent organism spent completely in freshwater and part of life cycle as an independent organism spent on land; fw/m = part of life cycle as an independent organism spent in freshwater and another part spent as independent organism in a marine environment.

3 Species Differences in Response to Toxic Substances: Shared Pathways of Toxicity — Value and Limitations of Omics Technologies to Elucidate Mechanism or Mode of Action

David L. Eaton, Evan P. Gallagher,
Michael L. Hooper, Dan Schlenk,
Patricia Schmeider, and Claudia Thompson

One of the major challenges in the field of toxicology is the dependence upon surrogate species as predictors of toxic response. Implicit in the value of cross-species extrapolations is the assumption that a chemical produces its adverse effects through a common mechanism. However, seldom does a chemical have only a single biochemical or molecular target within a cell, and many toxicokinetic and toxico-dynamic events within an organism, tissue, or cell can greatly alter the disposition of a toxic substance prior to its ultimate interaction with a molecular target. Omics technologies hold the promise of being able to capture an integrated snapshot of the entire biology of response to a toxic substance (at least in a given tissue) at a given dose and a given time point. Comparisons of such snapshots across species could yield great insights into shared pathways of toxicity, identification of potential molecular biomarkers of effects or susceptibility, and identification of important differences in response that could greatly affect how surrogate species are used in the risk assessment process.

There are numerous ways in which 1 species may respond quite differently from another when confronted with the same dose of a particular chemical. In some instances, differences in susceptibility may be due to differences in the structure and function of a single gene, although in other instances there may be multiple genetic differences that act in concert to produce a species difference. The list in Table 3.1

TABLE 3.1
Common pathways important
in conferring species differences
in response to xenobiotics

Differences in xenobiotic disposition

1) Absorption
2) Distribution
3) Metabolism (biotransformation)
4) Excretion

Differences in repair, regenerative,
or adaptive capacities

1) DNA repair
2) Tissue regeneration
3) Response to "oxidative stress"

Differences in molecular receptors

1) Endogenous hormone receptors
2) Neuroreceptors
3) Nutrient and cofactor receptors
4) Cell surface receptors
5) Nuclear translocators
6) Membrane transporters

Differences in signal transduction pathways

1) Apoptotis and cell cycle control pathways
2) Nuclear translocation
3) Immune responses and cytokines

---

provides a brief summary of the most common molecular and biochemical events that contribute to species differences in response to xenobiotics.

With this brief introduction, our working group set out to address a series of questions that pertain to how recent technological developments in molecular biology (so-called "omics" technologies) might be used to facilitate the mechanistic understanding of toxic responses, and why such responses might be conserved, or differ, among different species.

3.1 WHAT OMICS APPROACHES WOULD BE OF GREATEST VALUE IN PREDICTIVE TOXICOLOGY THAT UTILIZES BIOLOGICALLY RELEVANT EFFECTS IN ORGANISMS OR THE ENVIRONMENT?

A suite of omics-based markers (e.g., characteristic profile of genes, proteins, or metabolites) thought to be indicative of a pathway of toxicity needs to be tied temporally, spatially, and mechanistically to a biologically relevant adverse effect in an organism. This may be accomplished by linking the omic profile to an in vitro response that has been linked to a relevant whole organism endpoint. These new techniques and technologies hold promise in helping establish linkages between initiating events in a toxicity pathway and toxic responses in the whole organism. Which of the various omics approaches hold the most promise with regard to prediction will likely change in concert with the pathways of toxicity and, ultimately, will need to be incorporated through a "systems biology" approach that utilizes omics data to move from pathways to complex networks (Ge et al. 2003; Begley and Samson 2004; Morel et al. 2004). The integration of the various omics technologies (i.e., transcriptomics, proteomics, and metabolomics) into ecotoxicology research has been dubbed "ecotoxicogenomics" (Snape et al. 2004).

Pathways of toxicity can be broadly grouped by the type of chemical reactivity that occurs to initiate biochemical perturbations, including

- nonspecific and nonreactive interactions (where partitioning of lipophilic chemical into membranes in sufficient quantity results in general perturbations of membranes, membrane channels, membrane protein structures, etc.);
- nonspecific, but reactive (e.g., hard electrophiles that are DNA reactive), to soft electrophiles (protein SH reactivity); and
- specific protein interactions (e.g., binding to specific receptors, targeting specific enzymes, blocking specific ion channels, etc.).

To the extent that chemical exposure is linked to a specific compensatory response (e.g., DNA-reactive chemicals inducing DNA repair mechanisms or sulfhydryl-reactive chemicals depleting glutathione (GSH)), a characteristic "signature" or profile of gene/protein expression may serve as a marker for chemicals causing toxicity through these pathways. For example, chemicals that produce adverse effects via altering hormonal status (so-called "endocrine disruptors") may elicit specific patterns of gene or protein expression indicative of activation (or inhibition) of a particular hormonal pathway (e.g., estrogen, androgen, thyroid hormone biosynthesis) (see Francois et al., 2003, and Rotchell and Ostrander, 2003, for reviews).

Chemicals that disrupt the mitochondrial proton gradient and uncouple electron transport are likely to be characterized with metabolomics techniques before compensatory transcriptomics or proteomics responses are detectable, and are potentially more sensitive and reflective of actual biological effects. Thus, metabolomics approaches would likely be of benefit in predictive toxicology in terms of biologically relevant effects via the measurement of important cellular components such as adenosine triphosphate (ATP) production, glutathione, reducing agents, thiol and redox status, and other molecules that directly reflect health of the cell or organism. These cellular metabolites are conserved across species, thus eliminating species-specific differences associated with interpretation of transcriptomics data. Obviously, the advantage of omics technologies over more traditional single-gene approaches is that patterns of change in transcripts or proteins may reveal shared pathways downstream from the initial event that are by nature critical to the ultimate toxic response.

A key challenge in using omics technologies in species' extrapolations is validating that a particular pattern of change is strongly correlated with an adverse response in the whole organism. Implicit in the assumption is that characteristic patterns of change in gene, protein, or metabolite content can be used to identify shared mechanisms of toxicity across different species so that the experimentor can, in fact, identify structurally and functionally similar genes, proteins, and metabolites across species. At the present time, there are significant challenges in constructing accurate gene ontology, orthology, and annotational relationships across species. There is often confusion in the literature with regard to gene nomenclature and orthology across relatively similar animal species such as rats and mice, although these shortcomings have been substantially reduced with sequencing of the rat and mouse genomes. In general, the further one is removed phylogenetically from the test species, the more difficult it will be to assign proper orthology of genes/proteins from the test species to the extrapolated species in question. However, it is not necessary to have identified the ultimate molecular target responsible for a particular adverse response in order to make a rational biological conclusion that 2 species are likely to share a particular adverse response from a given environmental stressor.

At present, there are 2 general ways in which omic technologies can potentially be used to predict biologically relevant cross-species extrapolations (Figure 3.1). The first is based on hypothesis generation and involves the identification of response signatures indicative of injury. For example, work conducted initially in yeast has demonstrated that DNA damage to an organism generally produces an upregulation of various DNA repair genes, as well as alterations in genes governing cell cycle control and apoptosis (Gasch et al. 2001; Begley and Samson 2004). This is likely an adaptive response in an effort to correct the increased level of DNA damage before mitosis or to trigger cell death in the event that DNA damage is extensive and likely to induce severe, unrepaired (or incorrectly repaired) damage to DNA (Elledge 1996). However, although a signature response for DNA damage in yeast is reasonably well described, the characteristic of a signature DNA damage response in mammalian cells using gene expression microarray technology is far less compelling.

An enormous challenge is in the area of gene annotation and orthology across species. Comparing 2 microarray studies of the same toxicant, but done on different platforms, in different cell systems, and under different conditions (dose, time, culture media, etc.) may not result in a recognizable signature, even if there is some underlying common mechanism of action. For example, Hamadeh et al. (2002) and Bae et al. (2002) sought to characterize the DNA damage response of human keratinocytes to inorganic arsenic. Bae and colleagues used an Ad12/SV-40 immortalized human keratinocytes (RHEK-1) and the Clontech Atlas Human Cancer 1.2 Array containing 1185 genes; the Hamadeh group (2002) used normal human keratinocytes (NHEK) in primary culture and the NIEHS ToxChip containing 1906 genes. (Note that in both of these experiments, relatively simple arrays containing fewer than 2000 gene targets were used. Compare this to the common 20,000 to 40,000 gene target arrays in wide use now).

A comparison of these 2 papers to identify a common pattern is nearly impossible because of differences in gene annotation or terminology and because there were large differences in the specific genes represented on each array. Even with that, the conclusions of the 2 groups were strikingly disparate. Bae et al. (2002) state: "Particularly interesting was the strong *induction* of multiple DNA repair proteins,

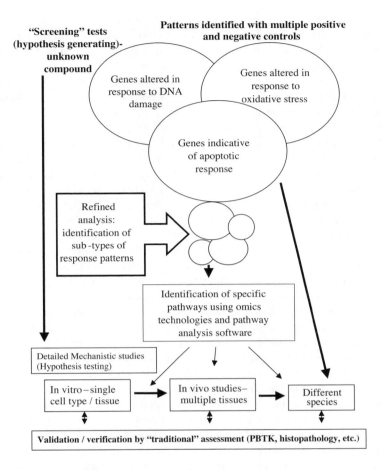

FIGURE 3.1 Use of omics tools for screening (hypothesis generating) and elucidation of potential shared pathways of toxicity across species.

including XRCC1, HNTH1, RAD23A and MLH1..." In contrast, Hamadeh et al. found that arsenic treatment resulted in a dramatic downregulation of DNA repair genes, including p53, damage-specific DNA binding protein 2 (DDB2), apurinic/apyrimidinic DNA lyase (APEX), DNA (cytosine-5-)methyltransferase 1 (DMT1), and hnRNP methyltransferase, S. cerevisae-like 2 (HMT1). Of course, these 2 studies used different doses of arsenic, different time points, different culture conditions, etc., so the results are not necessarily conflicting.

In 1999 the International Life Sciences Institute Health and Environmental Sciences Institute (ILSI/HESI) sponsored a collaborative scientific program that explored many of the challenges and opportunities in toxicogenomics. A diverse group of scientists from industry, government, and academia designed, conducted, and analyzed toxicogenomics experiments focused on identifying patterns of gene expression changes that would be reflective or predictive of various toxic endpoints, including hepatotoxicity, nephrotoxicity, and genotoxicity. The overall conclusions from the ILSI/HESI effort demonstrated that classification of the toxicity of compounds by gene expression profiling has potential merit, but also highlighted many of the technical challenges associated with transcriptomics. A full description of this large collaborative project was published in *Environmental Health Perspectives* (Pennie et al. 2004). Collectively, these studies point to the challenge of identifying true signature responses in gene expression for a particular agent (e.g., arsenic) or a particular type of global response (e.g., DNA damage).

Many other studies have also attempted to identify signature changes in gene expression in human cells that are reflective of a characteristic DNA damage response. In discussing the potential value of using microarrays to identify signature profiles for specific types of toxic responses, Burczynski et al. (2000) note that "Definitive demonstration is still lacking for such specific 'genetic fingerprints,' as opposed to nonspecific general stress responses that may be indistinguishable between compounds and therefore not suitable as probes of toxic mechanisms." To explore the potential of microarrays to identify predictive patterns of gene expression that can detect and distinguish toxicants with different mechanisms of action, they utilized HepG2 cells to evaluate the effects of 100 different toxic chemicals, using a small array representing 250 different genes. Cluster analysis of their results did identify compounds classified as DNA damaging agents and cytotoxic nonsteroidal antiinflammatory drugs (NSAIDS). Some characteristic changes in gene expression indicative of oxidative stress and apoptosis were also identified (Table 3.2).

Even this small array presented substantial statistical challenges in interpretation, which of course increase many fold with current arrays that contain over 40,000 gene probes. The problem is compounded further in cross-species comparisons because of the frequent uncertain nature of gene annotation, homology, and orthology across species. However, "streamlined" approaches for evaluating such putative signature pathways are available that greatly reduce the complexity of data analysis by simply limiting the number of probes on an array to signature target genes. For example, SuperArray Bioscience offers a small array containing oligo probes for 133 human genes identified as associated with DNA damage response, including genes involved in apoptosis, cell cycle regulation, and DNA repair (www.superarray. com/gene_array_product/).

Experiments establishing dose-related relationships in chemically mediated gene expression might further establish quantitative differences in sensitivity between 2 species. Although such information is of value in understanding a potentially important biological response, the information does little to identify the actual mechanism responsible for the species difference. It does, however, generate a series of hypotheses as to the mechanism of action, and it may be useful in identifying particularly sensitive species. If it is easier to measure DNA damage in each species directly than to evaluate the transcriptomics, proteomics, or metabolomics profiles, then little is actually gained by such experiments. A caveat would be if there is incentive to generate and test hypotheses associated with elucidation of the mechanisms underlying observed species' differences in susceptibility. Of course, the relevance of transcript profiles in predicting species responses depends upon some continuity in

TABLE 3.2
Examples of generalized adverse responses that might exhibit signature gene expression and/or proteomics profiles

Type of damage	Species, tissue, stressor	Responsive genes and/or pathways	Ref.
Oxidative stress	HepG2 cells, cisplatin	Mn^{++}SOD, epoxide hydrolase, glutathione peroxidase, glutathione synthetase, iNOS (inducible nitric oxide synthetase), aldehyde dehydrogenases (ALDH1 and ALDH2), CYP b5	Burczynski et al. 2000
	Sprague Dawley rat, liver, diquat	MLP2, NOS2, NADPH quinone reductase, interleukin 1 receptor	Waring et al. 2001
	60 various human tumor cell lines (not specified), artemisinins and anthracyclines	B-cell receptor-associated protein 29, CYP2C8, FMO1, GPX4, 3-hydroxyanthranline 3,4-dioxygenase, cytochrome c oxidase subunit VIa, COX6A1, ceruloplasmin, coporphyrinogen oxidase, CYP19A, CYP2B7, cytosolic epoxide hydrolase 2, ferridoxin I, FMO5, glutaredoxin 2, hGSTA2, hGSTM5, hGSTZ2, mGST2, mGST3, hypoxia inducible factor 1, catalase, aldehyde oxidase 1, NOS2A, NADH dehydrogenase 1α, nuclear factorI/C, thioredoxin I reductase, peroxiredoxin 1	Efferth and Oesch 2004
	PC12-D2R cells, H$_2$O$_2$	EGR1, MKP1, *c-jun*, *pc3*, core promoter element binding protein (cppeb), *c-fos*, *RhoB*, hsp70	Nair et al. 2004
Apoptosis	Rat, mammary tissue, organoselenides	↑ p21, APO-1, Caspase-3 ↓ cyclins, PCNA, *c-myc*	El-Bayoumy et al. 2003
	Human, TK6, BPDE	↑ Gadd45, Clusterin ↓ BCL-xL, *c-myc*	Akerman et al. 2004
	Human, HepG2, cisplatin Human, HepG2, BaP, DBA, carboplatin, mitomycin C	↑ p21, Fas, Bak ↑ Bax, Gadd45 PCNA, p21	Burczynski et al. 2000 van Delft et al. 2004

response within the genome of a species. Polymorphism and haplotype differences within a species could be as important as differences between species.

The second approach to utilizing omics technologies to infer a common mechanism across species is based upon hypothesis testing. In this example, omics tools are used to test hypotheses that a specific molecular pathway or biochemical network is linked to the development of the toxic response. Even at this level, omics tools may not necessarily identify the ultimate target for the initiating toxicological event. The latter will likely depend upon comparative analyses that utilize traditional biochemical and molecular approaches and that evaluate the functions of single genes and their products in the 2 species.

A species difference in response to a xenobiotic can be due to subtle differences in function between 2 highly homologous proteins, and even a detailed analysis of the transcriptome or proteome of the 2 species would not identify the ultimate difference. For example, a genetic difference in a gene encoding for a particular biotransformation enzyme that leads to a large difference in catalytic efficiency or turnover of a xenobiotic might lead to marked differences in the rate of activation or detoxification of the chemical among the 2 species, even though the level of gene transcript and protein expression are similar in both species. Likewise, small differences in structure of a receptor could lead to large differences in ligand binding efficiency, even if the level of expression (transcript and protein) were similar. In this regard, metabolomics approaches that identify changes in small molecules directly related to the function of the altered gene or protein could be of greater value in identifying the specific mechanism responsible for species differences in response.

3.2   HOW CAN OMICS BE UTILIZED TO UNDERSTAND MECHANISM AND MODE OF ACTION?

Initial considerations with regard to utilization of omics technologies to understand mode of action across species must begin with issues centering around hypothesis generation and testing. For example, if we assume that an organism is exposed to a toxicant known to bioaccumulate and produce structural changes (i.e., a histological lesion) in a particular tissue, and an omics profile is obtained from that tissue that represents a certain biochemical pathway, then a chemical target upstream of the biochemical pathway might be hypothesized. The chemical is tested and shown to interact with the cellular target, thus confirming the hypothesized mechanism. Potentially, a logical subsequent step would be to conduct experiments in another species of interest, and if similar biochemical or histopathological lesions were observed in similar target tissues, transcriptome-, proteome-, or metabolome-based profiles could be employed.

Using acute toxicity as well as chronic toxicity approaches, biologically significant points can be predicted based upon quantitative structure activity relationships as well as modes of chemical action. An example is the current research emphasis directed towards the linkages among modes of chemical action that may ultimately result in endocrine dysfunction (Francois et al. 2003). Ultimately, modes of action that are linked to reproductive injury can form the basis for additional quantitative structure activity relationships for untested chemicals. Additionally, the mode of action for many electrophile and proelectrophile chemicals can be identified from structural information, but only a few of these many reactive mechanisms have been characterized to the level necessary to develop quantitative structure–activity relationships for potency predictions. The new omics technologies could then be used in a hypothesis generation approach to better discriminate modes or mechanisms of action for other chemicals.

3.2.1 Discriminate between Defense and Adaptive Mechanisms from Direct "Toxic Response" and Secondary Downstream Events Responsible for Pathology

The ability to detect reproducible gene expression patterns that are consistent with a class of toxicants, but are also different across chemical classes, is a key factor in discriminating adaptive and toxic responses from secondary effects related to cell injury. Inherent in this assumption is that changes in gene expression are causally related to toxicity or are temporally downstream from toxic exposure. Clusters of positive correlations can be observed for compounds that act upon mechanisms such as DNA damage or peroxisomal proliferation (Baker et al. 2004). Accordingly, quantitative analysis of gene expression profiles and attendant gene ontologies and pathway analyses should inherently allow recognition of signature patterns representative of specific toxicities. Once these gene expression patterns are characterized, the patterns could then be used to evaluate environmental chemicals or unknown compounds possessing undefined toxicity. Such an approach would allow creation of databases containing response patterns to various toxicants under standardized conditions. By comparing an investigational compound's chemogenomic profile with known signatures of toxicity derived from databases, it could be possible to match candidate profiles against known toxicity profiles and compare compounds by degree of toxicity.

In the example of a DNA alkylating carcinogen, it would be important to establish whether there is a signature of transcriptional responses in the organism that may identify those species more likely to be sensitive to carcinogenesis from those resistant species. Conceivably, a DNA damage gene signature profile could be used to discriminate those more sensitive species from the more resistant species, but as discussed previously, there remain many challenges in characterizing such signature profiles. For example, even if a pattern of gene expression is reflective of changes in DNA repair suggestive of increased DNA damage, it is uncertain whether such changes may actually be protective by enhancing the repair of the DNA damage. Alternatively, increases in error-prone repair pathways could lead to enhanced mutagenesis. In such an approach, potential target genes involved in the mechanisms of BaP carcinogenesis would be included in the gene signature patterns, including genes important in apoptosis (bcl-2, bax, caspase-3) as well as genes involved in cell cycle regulation (induction of p53) and DNA repair.

Genes associated with secondary pathologies and nonspecific injury would have to be identified. Such examples may include genes associated with matrix remodeling and blood vessel formation. Dose–response studies would be conducted prior to toxicogenomics experiments so that subtoxic doses of chemicals could be employed in toxicogenomics experiments to yield gene expression changes associated with adaptive responses or initial molecular mechanisms of injury, as opposed to changes associated with cell or tissue damage. Validation studies could be conducted with time course characterization of a subset of genes whose altered expression has been associated with cellular protection (e.g., metallothioneins, glutathione biosynthetic genes, glutathione transferases, transcription factors, chaperone proteins, etc.).

One approach would be to carefully establish a dose–response relationship that would establish an ED_{30} at the minimum number of time points that have been associated with changes in gene expression related to mechanisms of injury (e.g., 24, 48, or 72 hours are reasonable time points based upon cell and in vivo studies in several mammalian and fish models). Presumably, the defense and adaptive mechanisms would be detectable at doses of the chemical that do not result in irreversible toxicity. Subsequent validation studies should address discrimination of time points at which bioaccumulation and a chemogenic signature profile are obtained in the tissue, as well as time points when depuration occurs without sustained or progression of tissue injury.

3.2.2 Integrate Omics with "Traditional" or Alternative Animal Models

As discussed, these approaches necessitate creating a reference compendium of mRNA signatures (chemogenic profile) that can be used to predict toxicity in nontraditional test organisms such as fish. The basic steps in the traditional models would involve 1) generating a toxicogenomics database base for the evaluation of novel compounds, 2) the establishment of quality-checked and normalized mRNA expression data, 3) the construction of a reference database with a number of well-known and well-characterized compounds, and 4) the selection of toxicological marker genes based on statistical methods. After validating and refining the reference compendium, the system is used to classify novel, uncharacterized compounds. Including related functional genomics data allows the characterization of gene function and regulatory mechanisms to determine the compounds' mechanisms of action at various toxicological endpoints.

In the case of DNA damaging agents, the mode of action may involve DNA binding of reactive intermediates followed by mutational events in key genes (such as protooncogenes and tumor suppressor genes), alterations in cell cycle regulation, and, ultimately, clonal expansion of initiated cells. For compounds such as PAHs, the available information suggests that the mechanisms of toxicity in aquatic organisms are similar to those observed in mammals (Maccubbin 1994). In the absence of a DNA repair damage response, it is unlikely that the chemical would act as a carcinogen, at least in the species tested. Again, laboratory studies would have to show induction or repression of genes related to these pathways in aquatic or alternative species (see following comments on the lack of studies demonstrating inducibility of DNA repair genes, cell cycle genes, and apoptosis regulatory genes in fish exposed to environmental agents).

If similar responses are observed across fish and rodents, then it is likely that the mode of action of these chemicals would be similar across other organisms. It is likely that relatively small numbers of genes may be sufficient to distinguish a variety of different toxic mechanisms, so targeted gene chips containing relatively few (100 to 250) genes could be used to demonstrate the appropriateness of the alternative models. However, it is fair to question whether such knowledge is more informative than more standard techniques for measuring DNA damage.

3.3 HOW DO WE INTEGRATE RESPONSES ACROSS GENE EXPRESSION, PROTEOMICS, AND METABOLOMICS AND APPLY THIS TO MAKE A SCIENCE-BASED STATEMENT ABOUT HEALTH OF AN ORGANISM?

Understanding the chemical mode of action (MOA) is a critical element for extrapolating toxic effects observed in one species to other species. Thus, to gain acceptance of omics tools (individually or in combination) as diagnostic of responses to toxicant insult, their correspondence to traditional measures or markers of MOA and resulting whole organism phenotype must be demonstrated. Once this is done for a given toxicity pathway within a species, the omics tools can be tested for ability to predict across species. To verify that an omics profile is truly representative of a toxic MOA and not just a signature for one chemical in 1 organism under 1 set of conditions, additional chemicals known to act through the previously defined toxicity pathway (positive controls) must be tested. This will serve to refine the omics profiles for a specific mechanism and also provide further analysis of the value of combined transcriptome, proteome, and metabolome profile in comparison to the diagnostic power of these techniques individually.

Additionally, to assess whether a transcriptomics, proteomics, or metabolomics profile is sufficient on its own to diagnose or discriminate pathways of toxicity or whether it can or should be integrated to provide a science-based statement about the health of an organism, the individual responses should be referenced to toxicant dose. It may be necessary to obtain tissue-specific dosimetry as well as whole-organism doses–responses dependent upon where a metabolomic profile is obtained in relation to the tissue or cell type source of the transcriptome and proteome response profiles.

Coupling metabolomics, transcriptomics, and protein expression information has shown promise in several areas of medicine, including drug safety evaluation and disease diagnosis and staging (Lindon et al. 2004). In this regard, identification of potential chemical biomarkers linking exposure and adverse effects could be obtained through evaluation of transcriptome, proteome, and endogenous metabolite profiles. With an understanding of the mechanism of action, critical proteins or transcripts related to DNA injury, apoptosis, and cell cycle regulation could be targeted. Metabolomics approaches could be used to identify key metabolites that would aid in assessing toxicant-induced injury in cellular components such as glutathione and other low molecular weight thiols that may be targets for electrophilic intermediates. For in vivo exposures, profiles demonstrating alterations in endogenous metabolites may be assigned to particular tissues and disease categories that can serve as the basis for extrapolation of function across species.

For example, individual metabolites have been identified for disease biomarkers. These include cholesterol lipid profiles associated with heart disease, elevated prostate specific antigens and prostate cancer, as well as glucose levels as indicators for diabetes. Clinical indicators such as serum ALAT, ASAT, blood urea nitrogen, and

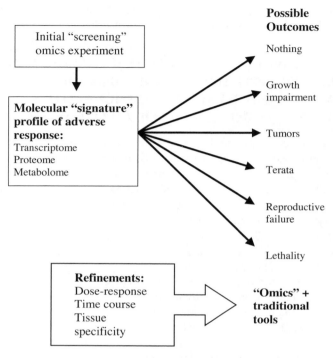

FIGURE 3.2 Importance of "anchoring" and validating omics data to actual toxicological endpoints, using additional omics evaluations and traditional approaches.

electrolyte profiles have been used in fish health diagnostics. These biochemical and endogenous metabolites could be exploited for metabolomics applications. Thus, metabolomics approaches currently provide links to clinical outcomes based on an organism's health in a more direct fashion than proteomics or transcriptomics. Because many known differences in chemical effects across species are due to differences in function even when gene or protein structure is largely homologous, metabolomics may initially hold promise for informing interspecies extrapolation of chemical toxicity. To the degree that transcriptomes and proteomes can be correlated to functional endpoints, the integrated profiles may gain better acceptance for species extrapolation. As previously mentioned, such an approach is strengthened by linkage to classic clinical and histopathology information (e.g., "phenotypic anchoring"; Figure 3.2).

3.4 HOW DOES DEVELOPMENT OF OMICS TECHNOLOGIES AFFECT THE INTERSPECIES EXTRAPOLATION PROCESS?

Interspecies extrapolation would benefit from omics technologies in 2 primary ways, depending on whether the technology is used for assessment of effects in exposed species or for the prediction of susceptibility in species of concern.

3.4.1 EFFECTS ASSESSMENT IN FIELD STUDIES

In this case, the focus is on species that are potentially exposed to toxicants in the environment and for whom there is a desire to understand the range of potential toxic responses. Transcriptome, proteome, or metabolome profiles could be measured as an indicator of the nature and severity of biochemical or pathological changes that occur in response to toxicant exposure. The monitored species would be sampled and assessed for characteristic patterns of response in appropriate tissues. Effective application of this approach would require sufficient understanding of the toxicant-, dose-, and duration-dependent characteristics of the utilized omics response for the compounds of interest. This necessitates that specific response patterns diagnostic of chemical toxicity pathways (e.g., signaling processes and responses at various levels of biological organization traditionally used to indicate mechanisms and modes of action for contaminant classes) are well documented and accepted.

The development of detailed surrogate species response patterns for chemicals of particular concern would provide a starting point for assessments of other species of similar taxonomic classification (e.g., birds, fish, mammals, plants). Decisions on which focal omics areas to pursue for a new species could be based on successful results of surrogate species tests. Confirmatory assessments would demonstrate the omics response similarity (and thus the utility) or dissimilarity (and need for more in-depth study) that a new species exhibits when compared to surrogate species. With time, a library of chemical toxic mode of action-specific response patterns could be developed that could help anticipate responses across a greater variety of species and chemical combinations. Such an approach would strengthen the extrapolation power based on subtaxa (i.e., raptors, passerines, waterfowl) and other life-history traits that might confer particular response patterns. Inclusive in this would be profile development that might help account for known species-specific sensitivity or resistance patterns.

The preceding approach would depend on an existing knowledge of omics responses diagnostic of toxicant modes of action of interest. In the case of field monitoring studies, samples would be required from unexposed individuals or individuals across a gradient of high to low exposure levels (if known) for determination of constitutive expression and magnitude of transcriptome, proteome, or metabolome response. A particular concern in areas of high species diversity would be the presence of species for which verified response profiles for the chemicals of interest are not available or for whom no molecular biology data exist at all. When extrapolation of omics approaches is made to a previously uncharacterized species, there must be some knowledge of variability in omics response patterns in characterized species to assess a level of confidence in the presumed similarity.

To circumvent the lack of established response profiles likely encountered for many species, application of methods that report uncharacterized transcripts and proteins in dose-dependent assessments in addition to identified transcripts and proteins might provide data on responsive components in the absence of detailed knowledge of gene source or protein identification. Subsequent sequencing of previously uncharacterized components could allow comparisons with known responsive components of surrogate or library species. Thus, insights into whether these independent observations are consistent with known patterns may be made or determinations that further testing is needed for this species may be possible.

Alternatively, given the current status of transcriptome and proteome technologies and their dependence on preapplication characterizations, metabolomics approaches that provide data on the metabolic status of studied animals may be more applicable to diagnosing chemical susceptibility for uncharacterized species because the endogenous metabolites are generally consistent across a wide variety of species and their response patterns lend themselves to interpretation without further characterization. Whichever approaches are utilized, development of methods that allow more efficient application of omics technologies to noncharacterized species is essential if these techniques are to be applied in cross-species assessments under true field conditions.

In the case of threatened or endangered species, application of omics approaches to circulating nucleated blood cells (fish) or feather pulps (birds) would provide a noninvasive means of sample collection for transcriptome, protein, and metabolic profiles, while urine could provide metabolic intermediates for analysis. In birds, care must be taken with this approach because their mixed fecal-urates could be contaminated with food-borne metabolic compounds (i.e., porphyrins) similar or identical to those under assessment. Combination of omics technologies with conventional biomarker methods could strengthen the utility and accuracy of both.

3.4.2 SUSCEPTIBILITY ASSESSMENT

Differential sensitivity to chemical toxicity is a function of the disposition of a chemical in a biological system and the sensitivity of the target molecules for the chemical's action. Furthermore, if the disposition or target sensitivity changes with chemical exposure or the exposed species can use alternative biochemical or physiological pathways to accomplish normal function, then these compensatory processes can decrease toxicity. Under this approach, an understanding of the disposition of a chemical, its direct interaction with a receptor and compensatory mechanisms leading to tolerance or resistance would be applied to predicting how a species would respond to an exposure of the toxicant. Availability of omics approaches to each of these areas is variable and eventual assessments will likely depend on conventional as well as molecular approaches, combined with chemical challenge experiments.

A chemical's disposition — its absorbance, distribution, metabolism and elimination in an organism — determines the dose and duration of chemical in the vicinity of critical receptors, the 2 principal determinants of the degree of chemical effect at the molecular level. Variability in the 4 dispositional factors can account for large differences in chemical sensitivity. Though many of the factors that determine disposition are a function of the chemical (i.e., lipophilicity, size, ionic nature, etc.), many other aspects of disposition are controlled by biochemical and physiological processes that can be assessed using molecular techniques. Active absorbance and elimination processes are dependent on transport proteins whose transcription message and structure could be compared between species. Distribution processes are heavily dependent on the interaction of chemicals with plasma, interstitial, and cellular binding or transport proteins whose transcriptome and proteome can be characterized under constitutive and intoxicated conditions.

The most important (and most studied) of the dispositional factors is xenobiotic metabolism: the conversion of chemicals into less toxic (detoxication) or more toxic (activation) biotransformation products. Substantial research efforts have been employed to characterize xenobiotic metabolizing enzymes such as the cytochrome P450 superfamily and the glutathione synthesis and transferase enzymes. Each of these 4 dispositional factors (absorption, distribution, metabolism, and excretion) combines to determine the level of exposure that occurs within an animal. A detailed understanding of the molecular processes that control their activity and expression can provide insight into a species' potential for accumulating a sufficient amount of a chemical to lead to toxicity. For any particular chemical of concern, species whose profile of dispositional factors (including transporters) was lacking critical protective mechanisms or had an overabundance of activating processes without concomitant increased detoxification would be considered to have greater potential for adverse effects when exposed to that chemical.

Besides dispositional factors, the sensitivity of molecular targets to chemical attack is the other major controlling factor in a species' sensitivity to a chemical exposure. Variability in the occurrence, structure, or availability of target molecules provides distinct differences in a chemical's ability to interact within a species and cause a toxic response. For example, the presence of the alpha2 mu globulin protein in male rodent urine makes them uniquely susceptible to the nephrotoxic effects of a variety of chemicals that bind to the protein. The absence of this protein in other species, such as humans, makes them essentially resistant to the effects seen in male rats (Swenberg 1993). The sensitivity of chemical targets depends on the structure of the target and how well it interacts with the chemical of interest.

Understanding the subtle target-specific structural requirements for chemical interactions would allow extrapolation to new, unstudied species through assessment of the structural characteristics of its receptor molecules. Substantial knowledge of receptor occurrence and sensitivity patterns has been generated in the absence of omics technologies; however, to the extent that this structure could be determined through proteomic methodologies, there could be a role for omics in the prediction of target sensitivity. The introduction of proteomic approaches to increase the diversity of species studied would be helpful in increasing libraries of species-specific baseline data for comparisons with uncharacterized species of interest.

There are many examples of how species can be protected by compensatory processes that decrease internal dose or confer tolerance to exposed animals. Dispositional changes that generally decrease internal chemical levels include induction of cytochrome P450 phase I and phase II enzymes, increases in glutathione production enzymes, and metallothionein synthesis in response to chemical exposures. These changes lead to increases in chemical oxidation and conjugation, electrophile sequestration, and metal binding, respectively. The transcriptome and proteome signatures for these enzyme or protein changes would be tangible targets for study, because their presence in exposed species of interest is predictive of capacity for metabolic compensation.

A change in receptor levels in the face of chemically induced overstimulation is another compensatory mechanism. If signals to downregulate receptors come via molecular pathways, these messages can be studied and their presence assessed as an indicator of a receptor-based compensatory mechanism. Finally, the availability of alternative biochemical or physiological pathways that bypass toxicant-induced damage plays an important roll in determining susceptibility. The utility of omics technologies to predict such pathways depends on their ability to measure critical components and endpoints of the pathways including their transcriptional message controls, protein constituents, and perhaps metabolic intermediates.

3.5 WHAT ARE KEY LIMITATIONS AND CONSIDERATIONS IN USING OMICS TECHNOLOGIES TO INFORM MECHANISMS OF CROSS-SPECIES DIFFERENCES IN RESPONSE TO XENOBIOTICS?

3.5.1 TIME OF SAMPLE COLLECTION

One of the greatest challenges to the design of omics experiments for the purposes of toxicological evaluation (including mechanistic assessment) relates to the time points at which the sample is collected for processing (RNA, protein, metabolite analysis). Because of the high expense and enormity of the amount of data obtained from a single experiment, seldom does one have the luxury of establishing a detailed time course of response. This is especially problematic for transcriptional analyses by microarrays because of the many downstream events that result in changes in transcriptional activity — including compensatory mechanisms — that occur in response to an initial insult. The nature of change in transcriptional profile in a sample collected 2 hours following an initial exposure, when compared to that observed at 4, 8, 24, or 72 hours following exposure, will be different. As some genes are activated, they in turn will trigger other changes in gene expression in a very dynamic manner, and the cell, tissue, or organism being examined will likely be undergoing constant fluctuations in gene expression as the tissue seeks to adapt to its perturbed environment. Thus, a "fingerprint" pattern for a particular change is not static, but ever changing over the time frame it takes for the cell, tissue, or organism to reach an adjusted steady state. This is also true for the proteome and metabolome, although arguably one might expect a somewhat "dampened" overall response in these parameters, relative to the rate and magnitude of change of the transcriptome in response to an external stressor.

Thus, in cross-species comparison studies, it is important that the time of sampling be carefully considered. For in vivo experiments, knowledge of the pharmacokinetics of the stressor in question in the 2 species would be of great value in ensuring that samples are collected at toxicologically comparable time points (e.g., at times when the maximal plasma concentration has been reached). For valid species comparisons to occur, a time course analysis (sampling and analysis at multiple times following exposure) would be most useful, but of course adds substantially to the expense of an experiment.

3.5.2 Duration of Exposure

Another chronology consideration in omics experiments relates to changes that occur following single-dose (acute) versus multiple-dose (subchronic, chronic) exposures. It is well established that adverse outcomes following high, single-dose exposures are usually dramatically different from adverse outcomes that occur following repeated, low-dose exposures. Utilizing omics data from a single, high-dose exposure to predict an adverse effect that might occur following low-dose, chronic exposure to the substances is of questionable value, unless one is able to demonstrate that a unique molecular event is unequivocally associated with the chronic response. For example, one might be able to conclude that a chemical that acts as a carcinogen via DNA adduct formation in a specific tissue is a carcinogen in one species, but not in another if profiles indicative of DNA damage are evident in the sensitive species, but not the resistant species, following a single high-dose exposure.

Likewise, if a chemical acts to produce a chronic adverse effect through a specific receptor, such as the peroxisomal proliferators activator receptor (PPAR) or an estrogen receptor, then marked species differences in changes in gene/protein/metabolite profiles responsive to the particular receptor would provide a reasonable indication of a species difference in susceptibility to the chronic effect of the chemical in question. However, whether such changes in mRNA, protein or metabolite profiles are adequately predictive of the adverse outcome so as to obviate the need for further toxicological evaluation depends heavily on the extent of validation of the association of the profile with the adverse effect.

3.5.3 Dose–Response Considerations

The genomics approach for mechanistic and predictive toxicology studies mandates that dose extrapolations can be made across species. As stated before, initial dose–response range-finding experiments would need to be conducted that establish a target dose (e.g., ED_{30}) that elicits a gene expression profile that can be linked with early-stage mechanistic and protective responses in the organism from those levels of exposure which elicit frank toxicity. Another approach would be to minimize the influence of potency differences among organisms by initial testing for cytotoxicity using conventional endpoints (e.g., 3-(4,5-dimethylthiazol-2-yl)-2,5-diphenyltetrazolium bromide (MTT) reduction or lactate dehydrogenase (LDH) release) in cell lines isolated from the different test organisms of interest (e.g., HepG2 cells from humans, hepatocytes from mouse, rats, fish, etc.). Of particular importance would be the identification of compounds that elicit unusual dose–response profiles. When considering systems such as the endocrine system, patterns of expression tend to be significantly different depending upon the stage of the affected cell or target. U-shaped response curves would significantly complicate interpretation and calibration of any omics endpoint, particularly those controlled through negative and positive feedback loops.

3.5.4 Target Tissues

Target organs may differ across species for some compounds. Organ function that is not conserved across species will be markedly different with regard to structure and cell types. For example, the teleost kidney differs markedly in structure and function from the mammalian kidney: Many fish have a kidney-like structure providing more of a hematopoietic function and the gill exists as the primary osmoregulatory organ. Furthermore, the teleost liver typically contains exocrine pancreatic cells that will yield considerably different basal transcriptome profiles than total RNA isolated from rat liver.

Many genes are regulated in a tissue-specific manner. Flavin-containing monooxygenase (FMO)1 is primarily expressed in human kidney and intestine, but in pigs it is the predominant hepatic form. With any given stressor, multiple systems and tissues may be affected, necessitating multiple tissue sampling and analyses. For example, if a compound affected hepatic and immune functions, liver and spleen/bone marrow, respectively, would have to be evaluated separately. Certainly, species differences in toxicant disposition and pharmacodynamics would influence the identification of target tissues and may explain species differences in response to any given toxicant.

Accordingly, transcriptome, proteome, and metabolome profiles will need to be coupled with knowledge of physiology and life history so that stronger predictions can be made with regard to mechanisms of toxicity.

3.5.5 AGE, GENDER

Within a species, the age and gender of individuals influence toxicity and must be considered when attempting to make cross-species comparisons. The age of an organism confers substantial variability to its physiology and constitutive biochemistry. Differences in the susceptibility of young organisms, compared to adults of the same species, can be due to underdevelopment of protective structures (blood–brain and blood–gonad barriers) and dispositional mechanisms (decreased metabolic enzymes, binding proteins and excretion processes, and increased GI absorbance mechanisms). Young are also more sensitive because of the presence of developing structural and functional processes whose disruption can have long-lasting effects that manifest in adult dysfunction. Also, the liver-to-body-weight ratio differs in young animals compared to old, and this can have significant effects on clearance. Older animals can also be more sensitive due to similar dispositional deficits as well as lifetime-accumulated nonlethal pathology.

Use of similarly aged individuals when performing cross-species assessments will avoid most age-dependent toxicity factors; however, their occurrence should not be dismissed and can present the most sensitive life stage in the organism's life. Gender-based sensitivity often occurs due to hormonal modulation of disposition. Oxidative metabolism enzymes can be highly dependent on hormone levels, leading to increased activation or detoxification in one sex compared to another. Renal transformation of chloroform to highly reactive phosgene occurs primarily in male rats due to testosterone-induced CYP2E1, which can be decreased with castration or induced in females with testosterone treatment. Because many of the causative factors in age and gender sensitivities are due to different patterns of biomolecule expression, there is a high probability that they could be studied effectively using omics technologies.

3.5.6 Nutrition

Gene and protein expression and metabolomics are heavily affected by nutritional inputs. Starvation or nutritive imbalance would dramatically alter profiles of each of the 3 omics methods. If an animal were sampled in the field under such conditions, much of the molecular profile observed would be associated with the animal's attempt to compensate for nutrient deficiencies, diminishing or masking the response due to chemical exposure. Species extrapolation would again be limited if animals at a specific stage undergo starvation as part of their life history (i.e., salmon, incubating birds).

For laboratory studies, test diets that have been carefully characterized for each test organism should be established. Of particular importance would be the control of antioxidants and other synthetic compounds (i.e., metals) that may affect susceptibility to toxicity as well as alter gene expression.

3.5.7 Conservation of Responses across Species (Induction or Repression)

Most of the investigations for toxicogenomics involve in vivo studies in rats because this species is commonly employed as the primary model by the pharmaceutical industry. Inherent in omics approaches is the assumption that the gene, protein, or metabolite signatures observed in nontraditional test organisms will be similar to those from studies using rodents or rodent cell lines. For these approaches to be of value with regard to interspecies evaluations, there is an assumption that some responses will occur in other species. Very little work has been done in this area with regard to inducibility of effects-based biomarker genes (GSH biosynthesis, apoptosis regulatory genes, DNA repair genes) in nonrodent species.

Basic studies of inducibility of DNA repair enzymes and other potential biomarkers of interest in the alternative models would need to be demonstrated. Carefully conducted correlational studies using quantitative polymerase chain reaction (PCR) analysis of targeted genes in the alternative model exposed to model compounds in the laboratory would be needed for these extrapolations. Laboratory studies validating responsiveness of key genes using nonrodent systems (in vivo exposures or isolated hepatocytes) and a targeted subset of potentially 20 to 50 key genes will be important. These genes may be those shown to be induced or repressed by other classical toxicants in other species (examples include startup CYP1A-induction by planar aromatic hydrocarbons, GST induction by synthetic antioxidants). In this regard, substantial work will need to be directed toward cloning, sequencing, expression, and characterization of the substrate specificity profile of gene products isolated from the alternative models.

3.5.8 Validation

Omics profiles that have been proposed to be indicative of specific modes of action or pathways of toxicity must be tested against multiple chemicals known to initiate the toxicity pathway in the same fashion, resulting in the same adverse outcome in the organism. Once an omics profile is verified as diagnostic using positive controls, it must also be shown to be unresponsive to negative controls (i.e., chemicals for which toxicity initiation and progression are "known" to occur via pathways other than that under study). It should then be tested in another species for which chemicals of the same class are known to produce the same whole-organism adverse effect, based on traditional measures. It would be important to include some of the traditional measures as well as dosimetry in the same or parallel studies to provide optimal linkage.

Gene expression results based on experiments with established class-specific toxicants (peroxisomal proliferators, alkylating agents, hepatic toxicants, etc.) should be clustered and compared to histopathology findings and clinical chemistry values. All results should show strong correlations among histopathology, clinical chemistry, and gene expression profiles induced by chemical agents. In addition, experiments should be conducted with investigators blinded to compounds.

Quantitative real-time-PCR (Q-RT-PCR) analysis should be conducted to analyze the expression of key gene transcripts whose expression appears to be altered on microarrays in response to chemical exposure. In addition, a subset of those genes whose expression does not differ among control and treated animals (excluding those housekeeping genes whose expression are generally unaltered by chemical exposure such as B-actin, GAPDH, as well as nonhousekeeping genes) should also be analyzed to ensure that false negatives are not generated. Experiments should be conducted using classical enzyme inducers (such as 3-MC, BNF, dexamethasone) and appropriate positive control genes whose expression is upregulated on exposure to the particular inducing agent, providing the basis for studies of corresponding functional variability in the gene products. Examples include the linearity of induction of CY1A and other Ah receptor response of genes by 3-MC. In general, the relative responsiveness of many of these genes will differ across species.

3.5.9 KINETICS, IDENTIFICATION OF RATE-LIMITING STEPS

Standardized reporting of dosing regimes associated with each omics response must be done to allow assessment of response in relation to disposition of the chemical, as well as its relevant biotransformation products, in select target tissues or the whole organism. Characterization of chemical kinetics, target organ dosimetry, and xenobiotic metabolism also provides a basis for responses (or lack thereof) to be properly interpreted across species.

Given the varied roles of enzymes and transporters in pharmacodynamic processes, understanding the relationships between multiple steps in any given mechanistic sequelae and the rate-limiting reactions is necessary. Thus, if induction of a phase II enzyme is observed following toxicant treatment without concurrent induction of the preceding phase I activation, the relevance of the phase II induction may be moot.

If a transcriptome or proteome response to a toxicant indicates involvement of a key enzyme (or enzymatic pathway) in the toxic sequelae, the enzyme's rate and capacity (kilometers and maximum volts) in relation to the toxic response must be confirmed by conventional enzyme kinetic measures. Ideally, this should be characterized for the toxicant (toxicant class) under study as well as standard substrates.

3.5.10 IN VITRO VERSUS IN VIVO STUDIES: CORRELATIONS

Omics tools can be used in in vitro and in vivo toxicology studies. In the case of in vitro models, changes in gene expression, protein patterns or metabolites can be identified in specific cell populations (cultured cell lines or freshly prepared cell isolates) or in complex mixtures of cells (isolated perfused organs, tissue slices, tissue cultures). The advantages of such approaches for comparative toxicology are obvious — for example, one could compare the response of hepatocytes isolated from rat, mouse, trout, bird, or human liver to a chemical stressor under identical conditions of dose and duration of exposure, and of sample collection and processing. Such comparisons could be extremely useful in assessing whether a given substance was likely to be comparably hepatotoxic in each species.

Although it would be tempting to infer that the results of such experiments could be directly extrapolated to the in vivo situation, there are of course many caveats to such conclusions. In the preceding example, it is assumed that hepatocytes isolated and exposed to the chemical in question in the in vitro environment function in a similar way in vivo. Species differences in the rate and/or extent of absorption, distribution, excretion, as well as possible differences in receptor affinities, adaptive responses, and any number of other physiological, pharmacological and environmental conditions, might completely invalidate the conclusions of predicted species differences based solely on the in vitro experiment. Nevertheless, such comparative in vitro approaches can be of value as screening tools to identify potential species differences in susceptibility, and they can facilitate the identification of mechanisms or critical pathways involved in the toxic response.

3.6 CONCLUSIONS

1) Omics technologies are powerful tools that, at the present time, allow generation of hypotheses likely to enhance elucidation of chemical mode and mechanism of action.

2) It is likely that gene, protein, and metabolite expression profiles can be generated for model environmental toxicants in a number of alternative models and that, through application of pattern recognition algorithms and computational analyses, patterns of expression will emerge that may be reflective of mechanism of toxicity. Others have demonstrated that chemicals from the same class of compounds give rise to discernible gene expression profiles that are strikingly similar to each other — more so than patterns corresponding to compounds from different classes of chemicals. Ultimately, relatively small databases based upon limited numbers of genes should be applicable for prediction mechanisms of toxicity based upon gene expression.

3) For a number of conserved biological pathways, it will be possible to generate molecular probes that are of value across species, and development of expanded databases for gene expression responses to environmental chemicals will likely be generated that will provide insights into pathways associated with adaptation, injury, and repair. For example, transcriptomics

and proteomics should be of value in identifying differences in species sensitivity to environmental stressors that act via endocrine-mediated pathways, via DNA damage, or through induction of oxidative stress.

4) Conversely, species differences that result because of pharmacokinetic or pharmacodynamic differences that are due to small differences in protein structure (changes in catalytic efficiency of an enzyme, alterations in ligand binding to a receptor) are not likely to be identified via current transcriptomics or proteomics approaches.

5) Omics technologies will likely be useful as screening tools to identify generalized modes of action, such as DNA damage, oxidative stress, and cell cycle disruption, as well as receptor-mediated adverse responses in nontarget organisms. However, before such screening approaches can be used to infer cross-species sensitivities, thorough validation studies are needed to firmly establish the association between patterns of change and a specific adverse outcome. This will require utilization of several positive and negative controls.

6) Although omics technologies hold great promise in helping to identify the mechanistic basis for predicting species similarities and differences in responses to xenobiotics, in most instances these tools will supplement, but not replace, routine toxicological testing and time-tested tools such as dispositional studies, clinical chemistry, histopathology, and issues relevant to life-history traits.

7) Transcriptomics is currently the most readily available approach to evaluating global changes in a cell tissue or organism in response to an environmental stressor. However, limitations in gene ontology, orthology, and annotation across species are major obstacles to cross-species comparisons using transcriptomics approaches.

8) Of the various omics technologies available, metabolomics holds the greatest promise in facilitating biologically relevant cross-species extrapolation based on shared pathways of biological function, particularly for species whose genomes and proteomes are not well characterized. However, current techniques for robust, comprehensive high-throughput analysis of the metabolome are in their infancy, and they require substantial gains in sensitivity, coverage, and cost before they will be of widespread use in cross-species extrapolation based on shared pathways of toxicity.

3.7 RECOMMENDATIONS

1) Laboratory studies carefully characterizing the inducibility or repression of genes governing susceptibility to toxicity need to be conducted in ecologically relevant, nonstandard species. Examples include analysis of genes involved in biotransformation, apoptosis, oxidative stress, and DNA repair. Studies are generally lacking that target characterization of the presence and functionality of *cis*-acting regulatory elements that confer inducibility of gene expression in alternative species. Furthermore, we

know relatively little of the transcription factors involved in the regulation of many of these genes or how those transcription factors may be regulated in nontraditional organisms.

2) Issues regarding false positives and negatives: Arrays developed on non-standard species should include validation studies ensuring specific hybridization of specific mRNAs, especially with regard to gene products of multigene families such as cytochrome P450s, glutathione transferases, uridinediphospho-glucuronosyl transferases (UDP-GTs), transporter proteins, etc. We cannot assume that a positive signal generated on an array from a surrogate species, even if closely related phylogenetically, encodes the target mRNA of interest.

3) Of consideration will be the availability and quality of arrays for target aquatic and terrestrial species. Arrays are now under development or are commercially available for zebrafish, largemouth bass, rainbow trout, and salmon. Certainly, limitations on the availability of arrays will help guide decisions related to selection of aquatic species for testing and also species extrapolations.

4) Dispositional studies conducted with target compounds in alternative species are needed to address pharmacological issues across phyla. Where possible, omics findings should be correlated to those known indicators of organism health in alternative models (e.g., electrolytes, histopathology, serum enzymes, electrolytes, thyroid profiles). However, clinical disease indicators have not been standardized with respect to fish and wildlife species. Studies to date indicate wide variances in clinical indicators of disease using veterinary and human medicine clinical parameters in feral organisms.

5) Robustness and applicability of primary cell models to in vivo findings involving the tissue of cell origin need to be established. For example, will comparisons of data generated from experiments targeting the effects of toxicant exposure on gene expression in primary rat hepatocytes lead to direct comparisons with effects observed in rat liver following treatments in vivo? Such studies will be crucial in determining whether transcriptional profiling in a model cell system is relevant to the in vivo scenario.

6) A key question in directing these approaches in fish and wildlife species with regard to elucidating mechanisms of action that occurs under environmental exposure scenarios centers upon whether the induction or repression of genes on an array will be sufficiently sensitive to predict changes that occur at relatively low environmentally relevant exposure concentrations.

7) With regard to nonstandard species such as fish and wildlife, the influence of diet, genetic heterogeneity, and environmental stressors (e.g., exceeding oxygen, temperature, salinity tolerances in aquatic organisms) will likely exert dynamic influences on mRNA and protein levels in these organisms and result in significant variability in gene, protein, or metabolite expression in relatively unexposed populations. Such variation could potentially mask changes strongly associated with exposure to chemicals unless such

exposures are of the capacity to produce relatively large changes in gene or protein expression.

8) Databases evaluating dose and duration of exposure at various life stages are necessary in surrogate species for which large toxicity databases exist.

9) Toxicants with known MOAs should be evaluated initially in species in which toxicity data are prevalent, developing "signature" responses. Subsequent to these studies, evaluations in additional species should be undertaken to assess species–species extrapolations.

10) Chemical MOA is a sound basis for extrapolating chemical toxicity across species. To gain acceptance of omics tools (individually or in combination) as diagnostic of responses to toxicant insult, their correspondence to traditional measures or markers of toxic MOA and resulting characteristic phenotype that represents a whole-organism adverse effect must be demonstrated. Currently, the utility of omics approaches to diagnose chemical MOA shows promise but has not been systematically tested for multiple chemical MOAs. The ability of these new tools to diagnose additional known toxic MOA should be evaluated through systematic testing in one species under standardized and well-documented test conditions. Once omics profiles characteristic of more that one toxic MOA are identified, and cross-chemical variability within an MOA sufficiently quantified and minimized, the ability of omics approaches to discriminate among toxic MOAs should be assessed.

11) The robustness of omics approaches to assign chemical mode of toxic action must first be verified within a species for a sufficient number of chemicals thought to work through a common toxicity pathway. Once this is done, the omics tools should be tested in another species known to be susceptible to chemicals operating through the same toxicity pathway.

REFERENCES

Akerman GS, Rosenzweig BA, Domon OE, McGarrity LJ, Blankenship LR, Tsai CA, Culp SJ, MacGregor JT, Sistare FD, Chen JJ, and others. 2004. Gene expression profiles and genetic damage in benzo(a)pyrene diol epoxide-exposed TK6 cells. Mut Res Fundamental Mol Mech Mutagenesis 549(1-2):43–64.

Bae DS, Hanneman WH, Yang RSH, Campain JA. 2002. Characterization of gene expression changes associated with MNNG, arsenic, or metal mixture treatment in human keratinocytes: application of cDNA microarray technology. Environ Health Perspect 110(Suppl 6):931–941.

Baker, VA, HM Harries, and others. 2004. Clofibrate-induced gene expression changes in rat liver: a cross-laboratory analysis using membrane cDNA arrays. Environ Health Perspect 112(4):428–438.

Begley TJ, Samson LD. 2004. Network responses to DNA damaging agents. DNA Repair (Amst) 3(8-9):1123–1132.

Burczynski, ME, McMillian M. 2000. Toxicogenomics-based discrimination of toxic mechanism in HepG2 human hepatoma cells. Toxicol Sci 58(2):399–415.

Efferth T, Oesch F. 2004. Oxidative stress response of tumor cells: microarray-based comparison between artemisinins and anthracyclines. Biochem Pharmacol 68(1):3–10.

El-Bayoumy K, Narayanan BA, Desai DH, Narayanan NK, Pittman B, Amin SG, Schwartz J, Nixon DW. 2003. Elucidation of molecular targets of mammary cancer chemoprevention in the rat by organoselenium compounds using cDNA microarray. Carcinogenesis 24(9):1505–1514.

Elledge SJ. 1996. Cell cycle checkpoints: preventing an identity crisis. Science 274(5293):1664–1672.

Francois E, Wang DY, Fulthorpe R, Liss SN, Edwards EA. 2003. DNA microarrays for detecting endocrine-disrupting compounds. Biotechnol Adv 22(1-2):17–26.

Gasch AP, Huang MX, Metzner S, Botstein D, Elledge SJ, Brown PO. 2001. Genomic expression responses to DNA-damaging agents and the regulatory role of the yeast ATR homolog Mec1p. Mol Biol Cell 12(10):2987–3003.

Ge H, Walhout AJ, Vidal M. 2003. Integrating 'omic' information: a bridge between genomics and systems biology. Trends Genet 19(10):551–560.

Hamadeh, HK, Trouba KJ, Amin RP, Afshan CA, Germolec D. 2002. Coordination of altered DNA repair and damage pathways in arsenite-exposed keratinocytes. Toxicol Sci 69(2):306–316.

Lindon JC, Holmes E, Bollard ME, Stanley EG, Nicholson JK. 2004. Metabonomics technologies and their applications in physiological monitoring, drug safety assessment and disease diagnosis. Biomarkers 9(1):1–31.

Maccubbin AE. 1994. DNA adduct analysis in fish: laboratory and field studies. In: Aquatic toxicology: molecular biochemical and cellular perspectives. DC Malins, GK Ostrander, editors. Boca Raton (FL): Lewis.

Morel NM, Holland JM, van der Greef J, Marple EW, Clish C, Loscalzo J, Naylor S. 2004. Primer on medical genomics. Part XIV: introduction to systems biology—a new approach to understanding disease and treatment. Mayo Clin Proc 79(5):651–658.

Nair VD, Yuen T, Olanow CW, Sealfon SC. 2004. Early single cell bifurcation of pro- and antiapoptotic states during oxidative stress. J Biol Chem 279(26):27494–27501.

Pennie W, Pettit SD, Lord PG. 2004. Toxicogenomics in risk assessment: an overview of an HESI collaborative research program. Environ Health Perspect 112(4):417–419.

Rotchell JM, Ostrander GK. 2003. Molecular markers of endocrine disruption in aquatic organisms. J Toxicol Environ Health B Crit Rev 6(5):453–496.

Snape JR, Maund SJ, Pickford DB, Hutchinson TH. 2004. Ecotoxicogenomics: the challenge of integrating genomics into aquatic and terrestrial ecotoxicology. Aquat Toxicol 67(2):143–154.

Swenberg J. 1993. Alpha 2u-globulin nephropathy: review of the cellular and molecular mechanisms involved and their implications for human risk assessment. Environ Health Perspect 101(Suppl 6):39–44.

van Delft JHM, van Agen E, van Breda SGJ, Herwijnen MH, Staal YCM, Kleinjans JCS. Discrimination of genotoxic from non-genotoxic carcinogens by gene expression profiling. Carcinogenesis 25(7):1265–1276.

Waring, JF, Jolly RA, Lum PY, Praestgaard JT, Morfitt DC, Buratto B, Roberts C, Schadt E, Ulrich RG. 2001 Clustering of hepatotoxins based on mechanism of toxicity using gene expression profiles. Toxicol Appl Pharmacol 175(1):28–42.

4 Bioinformatic Approaches and Computational Models for Data Integration and Cross-Species Extrapolation in the Postgenomic Era

Kenneth S. Ramos, Renae L. Malek, John Quakenbush, Ilya Shmulevich, Joshua Stuart, and Michael Waters

4.1 INTRODUCTION

The completion of the draft genome sequences for human, mouse, and rat, along with the development of genomics resources in a wide range of other species, has set the stage for the evolution of toxicology and environmental science from the study of 1 gene, 1 protein, or 1 metabolite at a time to more comprehensive approaches that allow profiling of the response of hundreds to tens of thousands of RNAs, proteins, or metabolites to external insults and stimuli. Fueled by the availability of fundamental data on the genes and proteins encoded within a wide range of species and motivated by the development of increasingly robust and reliable technologies for high-throughput analysis, new approaches in genomics, transcriptomics, proteomics, and metabolomics are beginning to provide data on a global scale. These approaches allow examination of changes in RNA, protein, or metabolite levels in the context of target genomes and the signaling and metabolic pathways that they encode.

Toxicologists were among the first to recognize the promise of genomics approaches for the understanding of fundamental biology. The emerging field of toxicogenomics represents concerted efforts to shed new light on understanding the molecular basis of toxicity and identifying compounds likely to trigger general or tissue-specific toxicity. The enthusiasm for toxicogenomics approaches is simple to understand: Toxicogenomics studies can provide information of the responses of tens of thousands of genes, or hundreds of proteins or metabolites, in a single assay. With such large quantities of data, it is now possible to extract patterns or biological relationships that can be used to elucidate the pathways involved in toxicity and to classify or prioritize chemicals for toxicity testing.

Despite the promise of "omics" technologies and the scientific advances made in recent years, many challenges to the application of genomics approaches in toxicology persist, including the ability to collect, manage, and analyze the data effectively, cross-species extrapolation, and data interpretation.

The earliest genomics approaches to examining gene expression were the use of DNA microarrays as a tool for probing transcriptional levels (Lipshutz et al. 1995). The research community quickly seized upon this approach as a means of identifying genes that might provide insight into a wide range of biological processes. Most of the early experiments adopted a simple design: comparing 2 biological conditions in order to identify genes that were differentially expressed between them. These experiments generally sought to gain insight into the underlying biology and used microarrays as a tool for identifying new genes that might contribute to phenotypic changes (DeRisi et al. 1996; Welford et al. 1998).

While making conclusive mechanistic inferences based on expression data has proven challenging, it quickly became clear that the utility of microarray-based approaches extended beyond mechanistic studies to functional genomics. Data from microarrays could be applied to a wide range of problems, including finding new subclasses within previously characterized states (Alon et al. 1999; Perou et al. 1999), identifying new biomarkers that could be associated with phenotype (Moch et al. 1999), and even using the expression patterns as biomarkers to distinguish phenotypic subclasses (Golub et al. 1999). In the analysis of gene expression in cancer, this has resulted in a proliferation of studies that have searched for patterns that could be used to classify tumor types (Sorlie et al. 2001) and to predict outcome (van de Vijver et al. 2002) and response to chemotherapy (van't Veer et al. 2002).

In the field of toxicology, omics technologies were quickly recognized as tools to investigate the responses of organisms to stressful stimuli. In its broadest sense, the science of toxicogenomics combines genetics, analysis of genomics-scale mRNA expression (transcriptomics), cell- and tissue-wide protein expression (proteomics), metabolite profiling (metabolomics), and bioinformatics with toxicology to understand the role of gene–environment interaction in disease (Ramos 2003). Early studies in toxicogenomics showed that classes of toxicants and toxic responses could be recognized by their gene expression signatures, recognizing that changes in mRNA levels may reflect primary alterations induced by the chemical or secondary responses to toxic insult.

Global gene expression profiles for chemicals representative of different modes of action were found to provide gene expression signatures of chemical exposures in male rats (Waring et al. 2001; Hamadeh, Bushel, Jayadev, Martin, et al. 2002; Hamadeh, Bushel, Jayadev, DiSorbo, et al. 2002). These studies were performed on acutely stressed animals, and the expression patterns appeared to be representative of the adaptive or pharmacologic activity of the chemicals. Enzymatic markers were used in conjunction with histopathology to facilitate "phenotypic anchoring" of gene expression data in conventional toxicological indices (Tennant 2002). By relating molecular expression to phenotypic anchors, it was possible to distinguish toxicological signals from gene or protein expression changes unrelated to toxicity. Phenotypic anchoring is a central step toward understanding the sequence of key events that begins to define a mode of action of a toxicant.

Waring et al. (2002) reported the use of a mode-of-action classification of a novel exploratory inhibitor of NF-κB-mediated expression of adhesion proteins, A-277249. A-277249 was known to cause liver hypertrophy and elevation of serum levels of alanine aminotransferase (ALT), aspartate aminotransferase (AST), and alkaline phosphatase (ALP) that serve as biomarkers for hepatic damage and necrosis (Travlos et al. 1996), although the mode of toxicity was not known when the study was initiated. Using a library of gene expression profiles of 15 known hepatotoxicants as a reference, A-277249 was reported to be most similar to Aroclor 1254 and 3-methylcholanthrene (3MC). In addition, several genes regulated by the Ah receptor were increased following exposure to A-277249, suggesting that, similarly to Aroclor 1254 and 3MC, A-277249 was acting through the Ah receptor.

Commercial vendors have developed extensive databases to make signatures of pharmacological and toxicological responses available to clients for use in drug development and safety assessment (cf. http://www.genelogic.com/, http://www.iconixpharm.com/). While a number of disease-related or toxicologically relevant gene and protein signatures have been reported in the literature, no public database has yet been established to contain and disseminate this information.

The field of toxicoproteomics came of age with the use of 2-D PAGE to characterize proteins from organisms treated with different toxicants. In the first of 2 studies, proteomics was used to dissect the mechanism of glomerular nephrotoxicity caused by puromycin (Cutler et al. 1999). In the second study, Fountoulakis et al. (2000) reported a database of proteins expressed in mouse liver and the changes in level of approximately one third of these proteins in response to a high dose of acetaminophen (APAP), but not in response to a similar dose of the relatively nontoxic isomer 3-acetamidophenol (AMAP). The experimental design of comparing a toxic and nontoxic isomer facilitated protein expression changes related to a class of chemicals to be distinguished from a response due to toxicity.

More recent experiments began to correlate gene expression profiles with other well-defined parameters, including pathological or physiological response phenotype, or other indices of toxicity. For example, experiments have been designed to correlate gene expression patterns with disease pathologies such as necrosis, apoptosis, fibrosis, or inflammation (Vilain et al. 2003; Hamadeh et al. 2004; Wagenaar et al. 2004). By relating molecular expression to phenotypic anchors, the goal is to distinguish toxicological signals from gene or protein expression changes unrelated to toxicity.

In combination, the global omics technologies can potentially provide a comprehensive integrated view of the function of the genetic and biochemical machinery of the cell. The application of these technologies to toxicology is based on the assumption that the sequence of events following toxicant exposures and leading to adverse events at the cellular and organism level will include critical changes in certain mRNAs, proteins, and metabolites. Consequently, monitoring these changes should provide insight into the molecular mechanisms of toxicity and, given a reference set of studies on prototypic toxicants, should be diagnostic for a given mode of toxicity. These relationships have been exemplified in several recent studies identifying novel molecular targets of toxic injury (Johnson et al. 2003).

The challenge moving forward is to extend the application of omics technologies to the study of toxicological response. For this to be achieved, a wide range of issues must be addressed, including understanding the limitations of these technologies and the need for replication, validation, and the associated effects of variability on the choice of sample size; our ability to extrapolate from experimental animal models to humans or other species; and the problem of assessing ecological impacts based on limited laboratory studies. In this chapter, we address this problem with a focus on computational aspects associated with large-scale studies.

4.2 MECHANISTIC VERSUS CLASSIFICATION STUDIES

Functional genomics studies are often treated as a single, monolithic approach to investigate a biological system. In fact, a wide range of technologies can be applied, even to the profiling of expression patterns of a single biomolecule within a single domain. RNA expression profiling using microarrays can involve single-color array platforms where each sample is independently assayed using an independent array, or 2-color platforms where query versus control (or other) samples are directly compared. The sets of probes used on the microarrays also varies, depending on application, and can include cDNA clones, "long" oligonucleotides (typically 50 to 70 mers), or short oligonucleotides (25 mers). Also, a wide range of microarray substrates, amplification and labeling protocols, detection techniques, image processing algorithms, and data analysis protocols — as well as other "large-scale" approaches such as quantitative real time-polymerase chain reaction (qRT-PCR) and serial analysis of gene expression (SAGE) — can be used in a similar fashion.

In proteomics, a range of approaches spanning protein–protein interaction applications to mass-spectral analysis of abundance is also available, as well as a host of separation techniques including liquid chromatography and 2-dimensional gels and mass-identification techniques. To date, however, proteomics experiments have been limited by the relatively small number of proteins that can be assessed at 1 time. In metabolic profiling, the protein-identification technologies are supplemented by NMR-based approaches to measuring metabolite profiles. While all of these distinctions are potentially important, in many respects they are details relative to the more fundamental questions being addressed using omics analysis.

Fundamentally, we can separate functional genomics experiments into one of 2 broad classes: mechanistic studies and class discovery or classification experiments. Mechanistic studies ask questions such as "What is the mode of action of a compound?" or "What gene products produce the different phenotypes observed?" These experiments are typically focused on understanding how genes and their products interact in response to a stimulus and ideally represent carefully crafted analysis of defined perturbations with sufficient biological replication to distinguish signal from noise. Such experiments can be very useful for delineating hypotheses to be validated in more focused studies' knockouts or phenotypic rescue to study the mechanistic associations derived from large-scale analyses.

In class discovery, a large number of samples are compared to ask whether the expression profile can be used to identify groups or subgroups in a collection of samples. Commonly used analytic approaches involve unsupervised clustering algorithms to reveal patterns in the data, but the challenge is to link newly discovered classes to meaningful biological endpoints, such as phenotype. Once classes have been identified, the next obvious question to ask is whether the expression profiles can be used as biomarkers to distinguish among the various classes. This typically involves use of a series of supervised statistical or other approaches to reduce the dataset to a minimal collection of genes (or proteins or metabolites) whose expression patterns provide a clear and reproducible distinction between groups.

Samples are assigned to particular biological classes based on some objective criteria. For example, one might examine 2 chemically related compounds, 1 toxic and the other nontoxic, to resolve patterns of gene expression in a microarray assay that can distinguish phenotypic outcomes. One of the first questions to consider is "Which genes best distinguish the various classes in the data?" The goal at this stage is to find those genes that are most informative for distinguishing the samples based on class. Fortunately, a wide variety of statistical tools can be brought to bear on this question, including t-tests (for 2 classes), analysis of variance (ANOVA; for 3 or more classes), and other supervised approaches discussed in Section 4.6. Essentially, each of these methods uses an original set of samples, or training set, to develop a rule for classification. These models are then applied to a new test sample and then use its expression vector sample to place the test sample into the context of the original sample set, thus identifying its class. The problem here, of course, is to understand whether the pattern observed can extend from the original test set to a more general set of samples — a problem in toxicology exacerbated by the challenge of linking profiles across species and the need to move from individual genes to biological processes or pathways.

While mechanistic and class discovery or classification applications may appear distinct, they need not be. One might, for example, use knowledge of a particular pathway to select a subset of the dataset to examine for evidence of phenotypic classes. Alternatively, the expression profiles most useful for distinguishing classes may be interpreted in order to formulate mechanistic hypotheses regarding the biological basis of the observed classes. While it is often difficult to provide mechanistic interpretation for signatures that are most useful for classification, this does not necessarily diminish the suitability of expression profiles for prognostic applications. It should be noted that many examples of biomarkers of unknown function, such as prostate serum antigen (PSA), are extremely useful as diagnostic or prognostic markers for various diseases. Whichever is the case, it may be more relevant to consider gene lists emerging from classification experiments as nothing more than sets of biomarkers with predictive applications. If the emerging biomarkers have a biological interpretation, this would indeed be a bonus.

4.3   COMPUTATIONAL METHODS FOR ORTHOLOGUE IDENTIFICATION

In order to realize the full potential of cross-species approaches to the analysis of toxicogenomics datasets, it is necessary to identify which genes are orthologous between organisms. Two genes are said to be orthologous if they were inherited by descent from a common ancestor and now reside in 2 separate organisms following a speciation event. Unless the biological and physiological roles of the genes have changed since divergence from the ancestral species, the genes will perform the same, or similar, functions in the derived species. Striking examples exist in which a gene from a human can functionally substitute for a gene in a model organism as distantly related as the fly. Therefore an accurate prediction of orthology is critical for transferring results from one organism to another.

Although orthologous sequences cover many different types of biological molecules, including proteins, RNAs, and *cis*-regulatory elements, the current focus of functional genomics studies and the primary focus of this section are to define methods that accurately predict orthologous protein-coding sequences. Protein orthologues are then used to relate genes from multiple species so that measurements collected from a gene in 1 organism can be related to data collected about the orthologous gene in a second organism. Predictive toxicology relies on extrapolating response from sentinel species to others and our ability to do this is predicated on understanding the relationships between the genes and pathways that are common across species.

The methods to attempt to predict protein orthology for eukaryotes accurately present several challenges, including the existence of numerous duplication and lateral transfer events, alternative and trans-splicing events, the presence of many proteins with multidomain architectures, and the incompleteness of genome sequence and its annotation, even for "completed" genomes such as human or mouse. A duplication event creates a copy of a gene referred to as a paralogue. While orthologous genes are related by speciation and are often assumed to retain the same function, paralogous genes are related by duplication and often diverge in function. As an example, consider the α-globin and β-globin genes in human and mouse. The orthologue of human α-globin is mouse α-globin and similarly for β-globin. However, α-globin and β-globin in human (or mouse) are paralogous genes because they are the result of an ancient duplication event. Orthologues exist between species, while paralogues exist within species.

It should also be noted that the age of a paralogue must be taken into account as well. Young paralogues — those arising from relatively recent duplication events — often retain the same function as the original gene, whereas ancient paralogues tend to diverge and acquire novel functions. When attempting to predict the orthologous relationships of duplicated genes, care must be taken to treat recent versus ancient paralogues differently. In fact, different kinds of paralogues can complicate genome-wide orthology predictions. Duplication events occurring before a speciation event produce so-called "out-paralogues." In such cases, it is important to match A with A′ and B with B′. Therefore, in our previous example, α-globin and β-globin are out-paralogues. Duplication events occurring after a speciation event produce so-called "in-paralogues." With respect to these 2 species, all 4 genes can be combined into the same orthology group because A and B are related by descent to A′ or B′.

The existence of paralogues in eukaryotic genomes presents challenges in predicting orthology in at least 2 ways: 1) They complicate the identification of true orthologues using sequence alignment-based strategies and 2) they increase a level of complexity into how orthologous relationships should be modeled and used to relate genes across genomes. In the first case, sequence alignment-based strategies (such as those employing BLAST) may have difficulty distinguishing true orthologues from paralogues. Because protein structure and, ultimately, protein function are selected by evolution, the true orthologue's sequence may differ to approximately the same degree as its paralogue's sequence has diverged from its ancestor. It may be that a protein only needs to conserve a relatively small domain to retain its function while the remainder of the protein can drift in sequence.

Paralogy also increases the complexity of relating orthologous sequences because of many-to-many relationships (i.e., groups of genes from one organism can be considered orthologous to groups of genes from another). In this case, it is not yet clear how datasets collected about genes that exist in complex orthologies can be transferred across species. However, this problem may be somewhat mitigated if we confine ourselves to core metabolic processes that are essential for organismal survival.

4.3.1 AVAILABLE ORTHOLOGY RESOURCES

Several orthology prediction sets are available. National Center for Biotechnology Information (NCBI) maintains the Clusters of Orthologous Groups (COG) database (http://www.ncbi.nlm.nih.gov/COG/), which associates genes into orthologous groups largely for bacterial genomes. NCBI's Homologene (http://www.ncbi.nlm. nih.gov/entrez/query.fcgi?db=homologene) and TIGR's Eukaryotic Gene Orthologues (EGOs) (http://www.tigr.org/tdb/tgi/ego/) provide a set of orthology predictions for eukaryotic organisms, but differ in the types of sequences used to construct the sets and the range of species represented. NCBI uses protein sequences primarily from the RefSeq collection, while TIGR uses their gene indices inferred from EST alignments. This allows EGOs to possibly associate different gene isoforms (alternative splice products) into separate groups. Additional orthology sets have also recently become available. For example, OrthoMCL (http://www.cbil.upenn.edu/gene-family/) also provides an orthology prediction for eukaryotic genomes.

The limitation of all of these attempts to define orthologues is that they are based on sequence alignments and the assumption of parsimony: that the smallest number of possible changes exists between orthologues. Deviations from these assumptions, incomplete gene catalogs, or other potential problems may cause misassociations between genes. Consequently, these candidate orthologues, while the most likely orthologues based on the available data, must be regarded in light of their potential limitations. The process by which most candidate orthologues are identified by most databases is described next.

4.3.2 ALL-AGAINST-ALL PAIR-WISE SEQUENCE ANALYSIS

Sequence homology searches are generally the first step in identifying candidate orthologues and the sequence alignment methods are used to determine how well primary amino acid sequences align. It is convenient to think about any significant alignment between 2 genes, A and B, as evidence for an orthologous relationship. Let A $\rightarrow$ B denote a putative orthology relationship found by querying protein A against a second organism's database within which B was found to be a significant match. Standard search and alignment algorithms like BLAST typically assign a score to the A $\rightarrow$ B match that reflects the significance of their alignment. The match score is a function of several parameters including the length of the protein sequences being compared, a scoring matrix that describes the similarities between different residues, and the resulting alignment that may include gaps and mismatches. Because these methods also typically incorporate the size of the database and the frequency of occurrence of sequence motifs in the target database, the A $\rightarrow$ B match will not in general have the same score as the B $\rightarrow$ A match because the genome size and composition between the organisms differ.

Most approaches for detecting orthologous genes first collect a comprehensive set of putative orthology links between 2 organisms. This is done by performing an all-against-all comparison in which each protein sequence from the first organism is aligned to every protein sequence in the second organism. Every significant A $\rightarrow$ B pair found from this step is then retained for subsequent analysis.

4.3.3 RECIPROCITY AND TRANSITIVITY

The simplest and most common postprocessing step from all-against-all comparisons is to enforce a notion of reciprocity. If B is A's best match and A is B's best match, then a best reciprocal hit A–B is said to exist. The reciprocity constraint groups genes A and B into the same orthologous group only if A–B exists. This is essentially a parsimony constraint: Two genes are most likely to be orthologues if they are the representatives in the 2 target species that have diverged the least. When 3 or more organisms are included in the analysis, the next most common step is to require an additional transitivity constraint among the best reciprocal hits. NCBI's COGs and TIGR's EGOs use a 3-way transitivity constraint (although EGO uses DNA rather than protein sequences to increase representation of genes and species). In these approaches, genes A, B, and C are associated together into the same group only if their best reciprocal links form a triangle spanning 3 different organisms — that is, the links A–B, B–C, and A–C must exist across all 3 species pairs.

4.3.4 PHYLOGENETICALLY BASED APPROACHES

Phylogenetically based approaches attempt to refine orthologous assignments by inferring an evolutionary tree relating a set of protein sequences using the previously defined phylogenetic relationships between the target species, based on other data, as a guide. Sequences identified to have diverged after a speciation event in the species tree are then predicted to be orthologues, while those that diverged before speciation events are predicted to be nonorthologous. One example of such an approach is resample inference of orthologues (RIO), which uses a bootstrap approach to learn a tree from a set of input sequences. In theory, these approaches promise to give a more accurate picture of orthology, but they are currently too slow for multiple genome-wide predictions of orthology (Zmasek and Eddy 2002).

4.3.5 FUTURE DIRECTIONS

The degree to which orthologous sequences have been altered since divergence depends on the type of biological sequence and the functions served within the system. Some proteins have a large range of mutational freedom and may contribute to the divergence between species while others, particularly those involved in key metabolic and other pathways, are highly conserved. For example, the human and mouse olfactory receptor genes are highly diverged between these 2 species, while elongation factor 1 alpha is highly conserved across species ranging from yeast to humans at the protein and DNA sequence levels.

It is likely that approaches based solely on protein sequence will be suboptimal. Data are now becoming available that will allow for more accurate orthology prediction. For example, true orthologues might also share conserved *cis*-regulatory elements and are likely to be positionally conserved relative to other genes within the genome. We might therefore be able to use the results from alignments of noncoding sequences between and around coding sequences for orthology prediction. Indeed, a number of resources such as VISTA (http://pipeline.lbl.gov) attempt to use genomics alignments to identify conserved regions and likely orthologues. Alternately, it may turn out that the correct approach is first to determine the set of conserved protein domains across sequence organisms and then identify which combinations are assembled in different species.

Although the existing computational methods are imperfect, they provide a first-pass catalog of likely orthologues. As the number of species we attempt to survey using genomics approaches grows, the increasing quantities of data we must analyze will likely require that we continue using automated approaches to identifying orthologous genes. While existing algorithms and datasets will undoubtedly continue to improve, we cannot forget the potential limitations that exist when interpreting data based on assumed orthology relationships.

As noted, a potential way to mitigate the effects of this problem would be to focus on core metabolic processes where the genes and proteins are much more likely to be conserved. Rather than looking to the response of individual genes, proteins, or metabolites, it is likely that the response of the pathways and networks that associate them has much greater predictive power than the individual genes alone, and that these evolutionarily conserved modules may indeed be the key to using orthologous relationships in predictive toxicology. For example, compounds that perturb cell cycle-related genes and pathways may lead to cancer in some species and apoptosis in others, but significant perturbations are likely to have deleterious outcomes nonetheless.

4.4   INTERPRETING EXPRESSION DATA ACROSS SPECIES

4.4.1   MOTIVATION

To study a diverse set of organisms using molecular approaches, it is important first to identify the core set of molecular pathways and processes shared across species. Gene expression data now available for multiple organisms are ideal for this purpose. DNA microarrays can give us a simultaneous observation on every gene in the genome for a single state of a cell. Expression databases contain many different regulation profiles showing how gene expression is perturbed by developmental stage, different growth conditions, stress, disease, and specific mutations. The large-scale datasets now available for gene expression give us an opportunity to explore which transcriptional changes play a significant role in toxicologically relevant processes. This section outlines 2 ways the analysis of gene expression data from multiple organisms is poised to play a significant role in toxicologically related processes. Algorithms that combine and interpret gene expression data across multiple organisms will allow us to

- identify core pathways shared across organisms,
- identify and characterize new genes involved in these core processes, and
- interpret gene expression patterns observed in one organism to another in terms of the up- or downregulation of these core processes.

The identification of core molecular pathways may inform the development of new technologies to monitor the health of a diverse collection of species at a higher resolution than is currently available. While many core processes have been discovered by classical approaches and are tabulated in databases like KEGG (htpp://www.genome.ad.jp/keg/)and GO (http://www.geneontology.org/), most of the genes in multicellular organisms still lack functional characterization. Even genes that have assigned annotations have been inferred from homologous sequences that have been characterized. It is therefore important to develop algorithms that can discover core modules from genome-wide functional approaches in a data-driven manner.

4.4.2   IDENTIFICATION OF CORE PROCESSES

Because genes that participate in the same pathway are often coregulated, it makes sense to search for core modules in the large databases of gene expression now available for multiple organisms. Computational methods that can identify conserved expression signatures from transcriptional networks on multiple species promise to shed light on core transcriptional processes common to a wide range of life. Development of algorithms that can use current gene expression datasets to elucidate conserved modules will therefore be critical for interpreting genome-wide toxicology datasets from a cross-species perspective.

Like many of the model organisms for which we have genome sequence, the genomes of organisms that are of toxicological and ecological interest often contain mostly uncharacterized genes. Thus, a major challenge for extrapolating data across organisms will be first to ascribe the function of genes for the majority of genes. A starting point for this approach is to relate uncharacterized genes to core transcriptional and protein modules.

The expression of genes under different conditions can provide information about gene function. Biological processes often involve the products of multiple genes acting in concert. For example, multiple genes can act together by encoding the proteins that make up a single complex. Multiple genes also can act together by encoding proteins that participate in a common signaling pathway. Therefore, gene expression measurements collected for a wide range of conditions over time for a single gene can be used to relate a gene to other genes in the genome. For this reason, compendiums of array data are often analyzed (Figure 4.1).

Computational methods exist that can ascribe functions for unknown genes if the unknown genes are sufficiently well correlated with genes or proteins of known function. One intriguing finding from large-scale gene expression studies is that genes belonging to the same functional group are often coregulated across a set of diverse conditions even when no specific experiments were designed to activate the functional group (Eisen et al. 1998). For example, genes and proteins known to be involved in muscle development and function were found to be significantly coregulated in a large collection of *Caenorhabditis elegans* experiments even though no muscle-related experiments were included in the dataset (Jiang et al. 2001). Thus, the coregulation relationships among the genes calculated over a large dataset can capture a wide range of anticipated and even unanticipated pathways. Similarity of expression profiles across diverse experimental conditions can suggest new relationships between genes.

However, coregulation detected in a single organism may not always imply 2 genes are functionally related. For example, *cis*-regulatory DNA motifs are predicted to occur by chance in the genome and might lead to serendipitous transcriptional

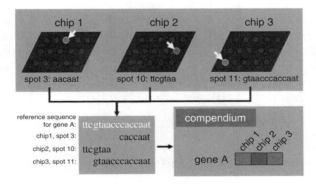

FIGURE 4.1 Analysis of gene expression measurements for a single gene under varying conditions can be used to generate array data compendiums to relate genes in the genome. If the expression pattern of gene A is significantly similar to that of gene B, the function of A may be inferred from the function of B.

regulation of nearby genes. Thus, it is likely that multiple protein–DNA interactions are frequently needed for transcriptional regulation, making the combinatorial probability of chance occurrence of the regulatory site probabilistically near zero within the genome. Using experiments limited to a single organism, it would be difficult or nearly impossible to distinguish coincidentally regulated genes from those that are biologically important. Recent evidence also suggests that seemingly unrelated genes can be significantly coexpressed if they are adjacent in the genome, possibly occupying shared chromatin domains (Spellman and Rubin 2002).

It is possible that the cell uses inexact mechanisms to regulate increases and decreases in mRNA levels. The presence of spurious coexpression in microarray data could confound gene function and core module prediction. Significant coexpression detected in a single organism will likely contain false-positive error; therefore, it is important to tease apart which coexpression relationships reflect the existence of true physiologically necessary links between genes and those that are spurious. Identifying these conserved links will improve core module prediction. This can be done in multiple ways: by using increasingly larger compendiums of microarray and other omics expression data to identify correlative expression patterns, by identifying syntenic relationships in the arrangement of regulatory mechanisms within the genome, or by directed manipulations of the expression of particular genes and other approaches.

Evolutionary conservation is a powerful criterion to identify genes that are functionally coupled from a set of coregulated genes. The recent availability of large sets of DNA microarray data for humans, mice, flies, worms, and yeast makes it possible to measure evolutionarily conserved coexpression on a genome-wide scale. Coregulation of a pair of genes over large evolutionary distances suggests that the coregulation confers a selective advantage — most likely because the genes are functionally related. Because small and subtle changes in fitness can confer selective advantage during evolution, the test for related gene function using evolutionary conservation in the wild is more sensitive than scoring the phenotype resulting from strong loss-of-function mutants in the laboratory.

Computational approaches that apply statistical techniques to detect significant coregulation across multiple species have recently been developed (Stuart et al. 2003; Ihmels et al. 2004). These studies have demonstrated that incorporating gene expression data across species gives a more accurate view of known core processes than can be obtained with the data from any single species alone. The natural corollary to these findings is that methods that incorporate expression across species will be more accurate for discovering new processes and relating genes to these processes in novel ways.

4.4.3 Using Core Processes to Interpret Gene Expression Studies

Investigators often characterize or identify several genes involved in a biological process of interest. They then want to be able to query the expression data to find new genes that are also likely to be involved in the process under study. Hypothesis-driven approaches are important, especially in the multiple-species setting where the context under which gene expression has been measured may not be known.

Given information about genes related in 1 organism gives us the ability to predict which treatments or responses measured across species might relate.

Current approaches could look for enrichment or depletion for core modules from a gene expression study. For example, if a set of genes is upregulated following addition of a toxin, one can identify the core modules that are affected. Standard approaches might use a hypergeometric overlap analysis to associate P-values with enriched modules. Similarly, a given module might contain genes that are differentially expressed up as well as down) in a given pattern of activities. If the occurrence of such patterns is more likely than by chance alone, we can conclude that the module is enriched (assuming of course, that we know the transcriptional responsiveness state of all the members of the module).

If we are given a set of related genes that have been isolated from genetic studies or identified in functional genomics screens, we can then use it as a training set proactively to find new candidate genes with similar expression characteristics. The availability of gene expression data across multiple organisms enables us to scan expression compendiums in multiple organisms to identify the combination of experimental conditions (e.g., different chemical treatments to different cell cultures) while at the same time identifying new candidate genes involved in the process.

4.5 INTEGRATING DATA ACROSS DOMAINS

In combination, the global omics technologies can provide a comprehensive integrated view of the genetic and biochemical machinery of the cell. The application of these 3 technologies to toxicology and environmental science is based on the assumption that the sequence of events following toxicant exposure and leading to adverse events at the cellular and organism level will include critical changes in certain mRNAs, proteins, and metabolites. Consequently, monitoring these changes should provide insight into the molecular mechanisms of toxicity and, given a reference set of studies on prototypic toxicants, should be diagnostic for a given mode of toxicity.

An integrated study of transcriptomics and proteomics responses to acetaminophen (APAP) at toxic and subtoxic doses was reported by Ruepp et al. (2002), including conventional clinical chemistry measures of ATP and glutathione in the target tissue as well as ALT and AST. This study yielded a mechanistic interpretation of the events, beginning with the formation of the toxic intermediates of APAP and leading to protein damage, glutathione depletion, and a hypothetical necrosis shunt in place of apoptosis due to loss of ATP. The responses of rodents to APAP are sufficiently well characterized to permit the use of these results as a check on the methodology, which recapitulated and expanded the biological responses expected from the mechanism of APAP toxicity.

The ILSI Collaborative Research Program to investigate the use of toxicogenomics in risk assessments came to fruition in 2004, with the publication of several papers reporting on the activities of 4 subcommittees. The committee found that microarray results from different laboratories and different platforms were comparable at the level of biological pattern, each identifying a common biological response, although the responses of the individual genes contributing to the pattern differed between platforms (Chu et al. 2004; Thompson et al. 2004; Waring et al. 2004). This finding supports the use of public toxicogenomics databases for meaningful meta-analysis of results obtained by different laboratories. As public repositories are developed, experiments will be deposited from disparate sources, utilizing different experimental designs, yet targeting the same toxicity endpoint or a similar chemical class. In these cases, it will be important for the repositories to have the capability to integrate data from different but related studies prior to performing data mining.

Based on the work of the ILSI Committee, the mechanisms of 2 hepatotoxicants, clofibrate and methapyrilene (Ulrich et al. 2004), and 3 nephrotoxicants, cisplatin, gentamicin, and puromycin (Kramer, Curtiss, S. et al. 2004), were reported. Additionally, applications of toxicogenomics to in vitro tests of genotoxicity were reported (Newton et al. 2004) and the results of an ongoing collaboration with the EBI established to develop a public database compatible with toxicogenomics endpoints (Mattes et al. 2004).

As the field of toxicogenomics matured, additional omics technologies were integrated into toxicogenomics studies. Experiments also began to focus on more challenging levels of organization, such as subcellular organelles (Jiang et al. 2004), nonstandard tissue such as saliva (Vitorino et al. 2004) or artery (Wesselman et al. 2004), less well-characterized species (Talamo et al. 2003), and genetic models of diseases (Lee 2004). Other important studies were exemplified by a report by Hamadeh et al. (2004) showing the integration of clinical and gene expression profiling to study effects of furan-mediated hepatotoxicity in rat liver.

4.5.1 Analysis of Multiple Domains' Omics Data

A key goal in toxicogenomics is to integrate data from different studies (Eckman et al. 2001) to produce a richer and biologically refined understanding of the toxicological response. It is anticipated that integrating data from different toxicogenomics domains will lead to synergistic interpretations beyond what can be resolved by analyzing the data in isolation. Examples include the interplay between levels of protein factors and gene expression changes, or between levels of metabolizing enzymes and the production or elimination of metabolites. The integration of data from different domains — for instance, proteomics and transcriptomics (Hogstrand et al. 2002; Ruepp et al. 2002) or transcriptomics and metabolomics (Coen et al. 2004) — has been reported, although in these studies omics data were analyzed from within the domain in which they were derived. The data were integrated at the summary level, following domain-specific analysis.

Cluster analysis and other unsupervised methods may be performed to derive global signatures of molecular expression from proteomics or metabolomics data in much the same way as with transcriptomics. For biological samples that segregate into unique clusters based on this analysis, additional efforts can be undertaken to discern novel proteins or metabolites responsible for the clustering, and further steps can be taken to evaluate them as potential biomarkers.

Extending and combining toxicogenomics with other approaches, such as physiologically based pharmacokinetic (PB/PK) and pharmacodynamic modeling, has been proposed (Waters, Boorman et al. 2003). PB/PK modeling could be used to derive quantitative estimates of target tissue dose at any time after treatment, thus creating the possibility to anchor molecular expression profiles in internal dose, as well as in time and phenotype. Relationships among gene, protein, and metabolite expression can then be described as a function of the applied dose of an agent and the ensuing kinetic and dynamic dose–response behaviors in various tissue compartments. Such models will (should) take into account the fact the transcriptome, proteome, and metabolome are dynamic systems subject to significant environmental influences, such as time of day and diet (Kita et al. 2002).

4.5.2 DEVELOPMENT OF A KNOWLEDGE-BASED SCIENCE OF TOXICOLOGY

The integration of data across domains has the potential to synergize our understanding of the relationship between toxicological outcomes and genome-level effects. In adopting an integrated approach to toxicogenomics, toxicology can progressively develop from studies predominantly on individual chemicals and stressors into a knowledge-based science in which experimental data are compiled and in which computational and informatics tools play a significant role in deriving a new understanding of toxicant-related disease (Tennant 2002).

Ideker et al. (2001) have used the phrase "systems biology" to describe the integrated study of biological systems at the molecular level involving perturbation of systems, monitoring molecular expression, integrating response data, and modeling the systems molecular structure and network function. In a similar manner, the term "systems toxicology" can be used to describe the toxicogenomics evaluation of biological systems involving perturbation by toxicants and stressors, monitoring molecular expression and conventional toxicological parameters, and iteratively integrating these data (Waters, Olden et al. 2003).

With the aid of the knowledge systematically generated and assembled (Zweiger 1999) through literature mining, comparative analysis, and iterative biological modeling of molecular expression datasets over dose and time, it will be possible to differentiate adaptive responses of biological systems from changes associated with or precedent to clinical or visible adverse effects. Ultimately, an integrated systems toxicology approach can be developed (Figure 4.2).

4.5.3 TOXICOGENOMICS DATABASES AND STANDARDS FOR EXCHANGE OF DATA

While several reports have described software for managing genomics/transcript profiling data at the local or laboratory level (Ermolaeva et al. 1998; Liao et al. 2000; Stoeckert et al. 2001; Bumm et al. 2002; Bushel et al. 2001), there are compelling requirements for public databases that combine profile data with associated biological and toxicological endpoints. Public warehouses provide a means for the scientific community to publish and share the data from experiments to advance understanding of biological systems. These repositories also serve as a resource for data mining and discovery of coexpression patterns. In a like fashion,

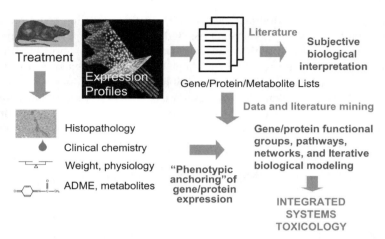

FIGURE 4.2 A framework for an integrated systems toxicology approach to study multi-species responses to chemical injury. A rat is shown as an example, but the principles apply to other species of toxicological and ecological relevance. Not depicted in the figure is the distinction between direct and indirect (adaptive) responses.

these repositories offer the regulatory community a reference resource for comparison with toxicogenomics data submitted as part of the compound registration process (Petricoin et al. 2002) and in evaluating classification algorithms and predictive models developed using proprietary datasets. Public toxicogenomics databases also offer the larger toxicology community common resources for comparing analytical tools and discussing experimental approaches. A database containing results from diverse laboratories and platforms would allow the identification and minimization of experimental practices that introduce undesirable variability into toxicogenomics datasets.

Another important function of public repositories is the promotion of international standards in data organization and nomenclature (Ball CA et al. 2002; Brazma et al. 2002, 2003; Edgar et al. 2002). Guidelines must be developed detailing what information must be included in a set of toxicogenomics data for it to be interpretable by others. In the domain of transcriptomics, minimum information about a microarray experiment (MIAME) guidelines (Brazma et al. 2001) allow sufficient and structured information to be recorded to correctly interpret and replicate the experiments or retrieve and analyze the data. Accordingly, guidelines for journal publication of microarray experiments have been proposed (Ball CA et al. 2002) describing requirements for submission of the data to either of the 2 existing public microarray repositories: ArrayExpress (Brazma et al. 2003) or GEO (Edgar et al. 2002).

Similar efforts have been initiated to prescribe minimal requirements for proteomics data capture and reporting a proteomics data integration model termed MIAPE (Orchard et al. 2004). Parallel efforts to standardize data submission and data capture standards for toxicological and toxicogenomics data have been launched by the CDISC/SEND Consortium and the National Center for Toxicology/EBI/ Health and Environmental Sciences Institute (HESI) collaborations. Of particular relevance is a recent proposal by the National Environmental Research Council to define MIAME requirements for the environmental genomics community (MIAME/Env). This proposal was incorporated into MGED in 2004 and, when fully implemented, will allow annotation of data to MIAME/Env standards and export as MAGE-ML.

Groups working towards data-exchange standards for toxicology and toxico-genomics include

- the National Institute of Environmental Health Sciences National Center for Toxicogenomics (http://www.nih.niehs.gov/nct), U.S.;
- EMBL-EBI European Bioinformatics Institute (http://www.ebi.ac.uk/ microarray/Projects/ilsi/index.html), U.K.;
- the International Life Sciences Institute's HESI (http://hesi.ilsi.org/ and http://hesi.ilsi.org/index.cfm?pubentityid=1), U.S.;
- Technical Committee on the Application of Genomics to Mechanism Based Risk Assessment; and
- the Microarray Gene Expression Data (MGED) Society (www.mged.org/).

and, more recently,

- the National Center for Toxicological Research (NCTR);
- Center for Toxicoinformatics, Food and Drug Administration (FDA);
- the CDISC/SEND Consortium (http://www.cdisc.org/); and
- the newly created MGED Toxicogenomics Working Group, part of the MGED RSBI (Reporting Structure for Biological Investigations) Working Groups (MGED, http://www.mged.org/).

In the field of proteomics, there are counterpart developments in the creation of the Human Proteome Organization's Protein Standard Initiative (PSI) protein inter-action format (Hermjakob et al. 2004). Similarly PEDRo prescribes standards for proteomic data for databases (Jones et al. 2004). A proteomics data integration model, MIAPE, represents minimum information about a proteomics experiment (Orchard et al. 2004), and the Standard Metabolomics Reporting Structure (SMRS) Group is developing metabolomics exchange standards.

Regulatory agencies aim to accept datasets with sufficient detail to permit accurate interpretation, without imposing unnecessary constraints on the consistency of terms used. Databases, on the other hand, need to map common terms in synonym lists to permit like data to be retrieved with a single query. This is done by first establishing a semicontrolled vocabulary, compatible with well-established tests and observations and also with the addition of novel tests unique to a given dataset. A second-generation toxicogenomics data-exchange standard can be envisioned to include a module to capture information about the study design and role of the individual subjects or animals as well as a toxicology-specific module to identify critical toxicity endpoints and other data. This new module should interface with MIAME-TOX as well as with MIAPE and metabolomics standards (for a toxicogenomics experiment encompassing these technologies) and be compatible with the toxicology data-exchange format being developed by the CDISC/SEND Consortium. An additional objective should be the development of standard data-exchange formats for toxicogenomics databases, including ontologies and data management software to ensure the accurate mapping of terms between different systems, so that different data systems can communicate effectively.

4.5.4 Systems Toxicology and Toxicogenomics Knowledge Bases

In order to develop a toxicogenomics knowledge base that will support the requirements of systems toxicology, we must address bioinformatics and interpretive challenges at multiple levels of biological organization (Figure 4.3). Our current level of understanding encompasses the lower levels of complexity (genes, gene groups, functional pathways) and our ability to describe and package that information in terms of our conventional understanding might be termed linear toxicoinformatics — that is, the description of stimuli and responses over dose and time following a toxicological stress.

Toxicologists and risk assessors typically define a sequence of key events and linear modes of action for environmental chemicals and drugs. In contrast, the networks and systems levels of biological organization reflect nonlinear cellular states in response to environmental stimuli. It must be recognized that the development of a knowledgebase to accurately reflect network-level molecular expression and to facilitate a systems-level biological interpretation is a complex issue requiring a new paradigm of data management, data integration, and computational modeling.

In systems biology, it has been proposed that structural biology, machine theory, and electronic circuitry may have enough in common to assemble the biological system from its parts (Nurse 2003). The full realization of this concept may take decades. However, toxicogenomics and systems toxicology models will not be assembled exclusively from knowledge of the architectural components of the cell

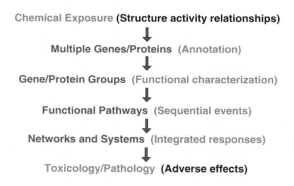

FIGURE 4.3 Interpretive challenges in toxicoinformatics at levels of increasing biological complexity.

without equivalent knowledge about the response of these components to perturbations (Begley et al. 2002). Thus, the "stress testing" of the structural biology of the system and the capture of those data in the context of the functioning organism adapting, surviving, or succumbing to the stress will be required.

Knowledge bases that fully embrace systems toxicology will contain precise sequence data that define macromolecules; interaction data based experimentally on colocalization, coexpression, and analyses of protein–protein interactions; and functional attribute and phenotype data based on knockouts, knockings, and RNA-interference studies, as well as studies of responses to chemical, physical, and biological stressors. These data will allow specific molecules to be accurately related to biological phenomena reflecting the normal as well as the stressed cell, tissue, organ, and system. Toxicogenomics knowledge bases must capture the detailed experimental and phenotypic information to document the interaction between informational molecules and structural moieties or modules within the cell as a function of dose, time, and the severity of the stress to the biological system. In the best of circumstances, a systems approach will build toxicogenomics understanding not only around global molecular expression but also around physiologically based pharmacokinetic and pharmacodynamic modeling and biologically based dose–response modeling.

4.5.4.1 The Chemical Effects in Biological Systems (CEBS) Knowledge Base

To promote a systems biology approach to understanding the biological effects of environmental chemicals and stressors across species, the Chemical Effects in Biological Systems (CEBS) knowledge base is being developed to house data from multiple complex data streams in a manner that will accommodate extensive and complex queries from users (Waters, Boorman et al. 2003). Unified data representation via a single modular systems biology object model (CEBS SysBio-OM) (Xirasagar et al. 2004) that incorporates current standards for data capture and exchange in multiple domains will greatly aid in integrating data storage and facilitate reuse of software to analyze and display data resulting from diverse differential expression or differential profile technologies.

Data streams include, but are not limited to, gene expression analysis, protein expression and protein–protein interaction analysis, and changes in low molecular weight metabolite levels, in addition to toxicology and histopathology data and pertinent information from the literature (Chaussabel and Sher 2002). It is envisioned that, as CEBS becomes a better utilized public resource, the data in CEBS will originate from disparate sources, utilizing different experimental designs, yet targeting the same toxicity endpoints or toxicant classes. In these cases, it will be important for CEBS to have the capability to integrate data from related studies and from the literature prior to performing data mining.

The conceptual design for CEBS is based upon functional genomics approaches that have been used successfully in analyzing yeast gene expression datasets (Hughes et al. 2000; Waters, Boorman et al. 2003; Waters, Olden, et al. 2003). Whole-chip sequence alignment of microarray probe or target sequence to the respective genomes within CEBS will facilitate the projection of expressed genes or proteins automatically onto known pathways (e.g., BioCarta, KEGG, GenMAPP). The knowledge base will support cross-domain (transcriptomics, proteomics, and metabolomics data integration with conventional toxicological and histopathological endpoints).

The similarities and differences in genomics responses among organisms in response to chemical and environmental exposures reflect fundamental interspecies evolutionary relationships. This interconnectivity can be tapped within a knowledge base such as CEBS to gain new understanding in toxicology as well as in basic biology and genetics. Advances in bioinformatics and mathematical modeling are expected to provide powerful approaches for elucidating the patterns of biological response embedded in genomics datasets. Thus, changes or differences in the integrated expression patterns of entire genomes at the levels of mRNA, protein, and metabolism can be determined. Collectively, these approaches greatly enhance our ability to address many of the major issues in human and environmental toxicology.

4.5.5 COMPARATIVE TOXICOGENOMICS

While there are important differences in the genomes and proteomes among species, many of the responses to chemical and physical stressors are evolutionarily conserved — for example, induction of xenobiotic metabolism, DNA repair, and apoptosis related to tissue regeneration. These conserved modules are important defensive responses common to many organisms, including humans, mammals, fish, birds, and invertebrates. By analogy to the GenBank, EMBL, or DDBJ databases for genome sequences, it will be possible to query CEBS globally, using a transcriptome or proteome of interest (or a list of outliers from expression analysis), to "BLAST" (Altschul et al. 1990) the knowledge base with a profile of interest and have it return information on similar profiles observed under other experimental conditions, including genes, groups of genes, metabolic and toxicological pathways and associated phenotypic information, and the dose, time, and phenotypic severity with which these effects are observed (Figure 4.4).

4.5.6 OPTIMIZING COLLECTION OF DATA AND DEVELOPMENT OF KNOWLEDGE

It is doubtful that attempts to extract toxicogenomics knowledge based only on the diverse datasets published in the open literature will be fruitful. In the case of transcriptomics, this is due in part to the fact that many of the published studies have been performed using ill-defined cDNA microarrays. In most cases, the probes on these arrays have not been subjected to requisite resequencing. Also, even under the best of circumstances, because of technical issues of interlaboratory reproducibility of array manufacture and differences in experimental design, these microarray experiments cannot be expected to yield exactly comparable data across laboratories. In general, only high-quality commercial oligonucleotide platforms provide sufficient reproducibility to permit careful cross-laboratories and cross-species investigation. Furthermore, cross-species investigations require the availability of comparable sequenced genomes.

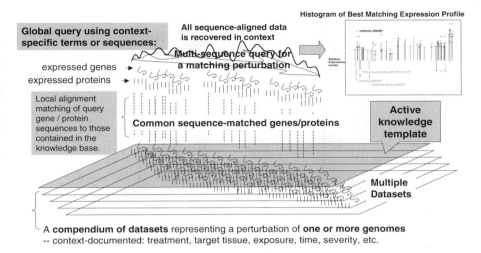

FIGURE 4.4 Sequence-based query of CEBS: A sequence-driven multispecies toxicogenomic knowledge base.

While not strictly computational, efforts at NCBI (NIH Intramural Sequencing Center (NISC) comparative sequencing program and elsewhere (particularly among commercial microarray manufacturers of oligonucleotide microarrays) are being undertaken to improve cross-platform, cross-laboratory, cross-species comparative analysis. Sequence alignment approaches (global and local) are being used by the NISC to identify orthologous gene relationships across species. Selective resequencing of sequenced genomes (for exon coding regions and intron noncoding regions) is being performed across genomes. Where exon sequence is conserved across species, similar function can be inferred. Sequence–alignment of orthologous genes expressed across model and toxicologically relevant species should be followed by the actual preparation and application of oligonucleotide arrays in functional analysis based on perturbation by known toxicants and stressors. These orthologous microarrays would be used to confirm putative gene function by direct experimental evaluation.

Alternatively, one can reconstruct orthologous relationships as illustrated in Figure 4.5. Such experiments should be performed using protocols that are as comparable as possible and data should be captured in a knowledge base that permits sequence alignment of microarray probes and proteins to genomes represented in the knowledge base and sequence-based queries across species.

In conclusion, the coordinated development of integrated cross-species, comparative multidomain investigations and the capture of the resulting data in a public knowledge base such as CEBS is recommended. Such cross-species studies of toxicity induced by drugs and environmental chemicals in sentinel and keystone species and extrapolated to human findings will provide support for qualitative health and environmental risk assessments. The combined and integrated data on gene/protein/metabolite changes collected in the context of genotype, dose, time, target tissue, and phenotypic severity across species will provide the interpretive information needed to define the molecular basis for chemical toxicity and to model the

Comparative Toxicogenomics

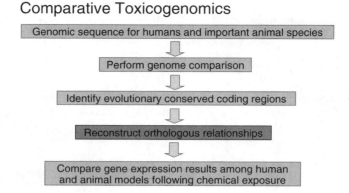

Modified from Thomas RE et al. (2003) Environ Health Perspect 110, 919-923.

FIGURE 4.5 Comparative toxicogenomics using transcriptomics.

resulting toxicological and pathological outcomes (Boorman et al. 2002). It should then be feasible to search for evidence of exposure or injury prior to any clinical or pathological manifestation, thus facilitating identification of early biomarkers of exposure, toxic injury, or susceptibility.

It is anticipated that toxicogenomics research will lead to the identification, measurement, and evaluation of more accurate, sensitive, and specific biomarkers that are ideally targeted to particular genetic subpopulations or disease types. In tandem, toxicogenomics research will identify molecules recognized as important factors in a sequence of events that will help define the way in which specific chemicals or environmental exposures cause disease. As such, toxicogenomics should help to delineate the mode of action of various classes of agents and the unique attributes of certain species and population subgroups that render them susceptible to toxicants as an important step in comparatively assessing potential human health and ecological risk.

4.6 SUPERVISED AND UNSUPERVISED ANALYSIS FOR TOXICOGENOMICS

DNA microarrays produce large sets of gene expression measurements often conducted to study relationships between biological samples. In this type of analysis, changes in relative gene expression for 2 different samples are often expressed as ratios. Because ratios are not symmetrical, a base-2 logarithmic transformation is applied so that the same absolute value is assigned to equivalent fold increases or decreases in gene expression. To analyze the complex datasets generated from microarray experiments, clustering algorithms are often used to identify potentially meaningful relationships.

Clustering refers to the process by which patterns or objects are classified into groups that share some measure of similarity and requires that a distance be defined to measure the similarity among objects within a group, along with a clear basis for partitioning these objects into groups. A general assumption underlying cluster analysis of gene expression data is that genes in the same pathway have similar expression patterns over a set of arrays (whether time-series or experimental conditions). The function of genes could be inferred on the basis of association where genes that appear in the same cluster are assumed to share similar functions or levels of coregulation. Time-series clustering is used to group genes whose expression levels exhibit similar behavior through time. Another application of cluster analysis is for grouping of samples (arrays) by relatedness in expression patterns. In the context of biomedical and environmental research, this approach allows the experimenter to discriminate pathologies or toxicity profiles based on patterns of gene expression.

One of the first efficient visualizations of microarray data was presented by Eisen et al. (1998), who used a matrix of colored cells where each column represented a sample and each row represented a gene and the brightness of the cell was proportional to the log of the ratio. In their colored matrix, red was used to represent upregulated genes (where the log of the ratio is positive), and green was used for downregulated genes (where the log of the ratio is negative). The image is then rearranged using a correlation metric such that genes are ordered so that those with similar profiles are placed together. This ordering of genes can be easily accomplished using clustering algorithms, generally based on intensity of expression or correlation of expression pattern. Clustering can also be applied to the samples to identify common patterns of gene signatures. Genes and samples are frequently sorted in a bidirectional fashion that allows for clear visualization of similarity in gene expression.

Supervised methods are used when the expression profiles can be assigned to one or more classes based on other biological characteristics. Thus, classification of gene expression data is often used to discriminate different types of samples, such as tissues, disease types, or treatments. The typical approach is to cluster genes by similarity in gene expression patterns and then utilize individual clusters that best discriminate the classes as markers. The use of known classes to build a classifier is termed "supervised clustering" and has been successfully used to classify cancer types and drug treatments. Supervised learning or classification can also be achieved using other established techniques, including weighted voting (Golub et al. 1999), artificial neural networks (Ball G et al. 2002), discriminant analysis (Amato et al. 2003), classification and regression trees (CART; Boulesteix et al. 2003), support vector machines (SVM; Furey et al. 2000; Ramaswamy et al. 2001), and k-nearest neighbors (kNN; Theilhaber et al. 2002), as well as a host of others.

Gene expression profiles refer to the expression values for a particular gene across various experimental conditions (or many genes under a single experimental condition). Considering the entire profile is a key analysis step toward revealing the function of genes (profiling) and discovering new classes of genes for classification (taxonomy). This type of analysis is limited by the large number of variables (or genes), the small amount of knowledge of the function for most of genes, and the lack of knowledge of the underlying classes or subclasses. Clustering algorithms differ on the basis of the manner in which they utilize the distances or similarities to group objects. In the case of hierarchical algorithms, clustering is based on successive splitting (divisive clustering) or merging (agglomerative clustering) of the groups, as defined by a measure of distance or similarity between objects. Another way to classify algorithms is based on their output: In hard clustering, the output is a partition of the data, while in soft (i.e., fuzzy) clustering the output is a membership function whereby an object (e.g., gene) can belong to different groups with different degrees of membership.

In hierarchical clustering, 3 common ways to update the distance between clusters are called "single," "complete," and "average linkage." In single linkage, when 2 clusters are joined into a new cluster, the distance between clusters is the minimum distance between any of the elements. In complete linkage, when 2 clusters are joined into a new cluster, the distance between clusters is the maximum distance between all the elements. In average linkage, when 2 clusters are joined into a new group, the distance is the average distance between the elements. Different linkages lead to differently shaped partitions, so the selection of the linkage must be determined by the type of data to be clustered and the outcome desired. For instance, complete and average linkages tend to build compact clusters, while single linkage tends to build clusters with shape more influenced by the data and thus more likely to be affected by spurious data points.

Proper selection of an algorithm is defined by the questions being asked. Each algorithm has its strengths and weakness and is better suited to a particular task. Hierarchical clustering algorithms are extremely powerful for exploratory data analysis because they do not need prior specification of the number of clusters, and their outputs can be visualized as a tree structure that is called a dendrogram. Most partitioning algorithms are based on minimization of an objective function that computes the quality of the clusters. Partitioning algorithms fail when the shape of the cluster is complex or dependent on groups of variables.

A broadly used iterative algorithm is the k-means algorithm, which is characterized by simplicity of implementation, high convergence speed, and good quality of clusters. The algorithm identifies a number (k or fewer) of clusters defined by centroids that minimize the distance among members of a group. In the k-means algorithm, each expression profile is classified as belonging to a unique cluster (hard cluster), and the centroids are updated based on the classified samples until the algorithm converges. In a variation of this approach, known as "fuzzy c-means," all profiles have a degree of probability of belonging to each cluster, and the respective centroids are calculated based on relative probabilities.

4.7 NETWORKS

Genes and their products in cells are not independent; rather, they exist in a highly interactive and dynamic regulatory network, regulated by rules that define the interactions between elements. However, discovering the network structure has thus far proven to be elusive because we lack sufficient information on the components of the network or lack the necessary multidisciplinary approaches that integrate biology and engineering principles and computational sophistication in modeling.

Our ability to elucidate the regulatory mechanisms in various organisms has the potential to make a substantial impact on our understanding of the features and properties that are ubiquitous across species. For example, we know that most living organisms have developed highly robust architectures to maintain cellular homeostasis under diverse environmental conditions and perturbations. Outcomes are reached by methods that are exceedingly parallel and extraordinarily integrated, as even a cursory examination of the wealth of controls associated with the intermediary metabolism network demonstrates. An important goal is to determine what features of the cell's control mechanisms, many of which may be reusable building blocks or motifs comprising specific interactions, are responsible for such homeostatic stability. For example, feedback regulatory circuits are routine even for the most common activities, resulting in a balance that favors survival of the organism in a complex environment. The intersection of the sets of motifs discovered in different organisms would likely point to generic mechanisms utilized across species. Biological networks, such as genetic or metabolic networks, can be conveniently studied from the perspective of complex systems.

A central tenet of the complex systems approach to biological systems is that the natural condition of an organism is the maintenance of a homeostatic state consistent with survival; as such, disease and other pathological or toxicological conditions perturb the system away from homeostasis. Our goal is then to understand the manner in which cells execute and control the enormous number of operations required for normal homeostasis, the ways in which cellular systems fail in disease, and the role of various molecular mechanisms in pathogenesis. Complex systems-based models of genetic networks are expected to help us consistently predict a cell's response to a stressful challenge and, in particular, to enable us to make such predictions across organisms and species. Such approaches inevitably require computational and formal methods to understand general principles governing the system under study and make useful predictions about system behavior in the presence of known conditions and perturbations. To this end, we need to examine a number of fundamental questions regarding the modeling of genetic regulatory networks and to identify key nodes in the network where modulation exerts the greatest disruption of normal physiology.

4.7.1 What Class of Models Should We Choose?

We would like to argue that this choice must be made in view of 1) the data requirements and 2) the goals of modeling and analysis. Indeed, data are required to infer the model parameters from observations in the physical world, while the model must serve some purpose — in particular, prediction of certain aspects of the system under study. Simply put, the type and quantity of data that we can gather, together with our prescribed purpose for using a model, should be the primary factors determining the choice of a particular model class.

However, any such choice involves a classical trade-off. A fine-scale model with many parameters may be able to capture detailed low-level phenomena such as protein concentrations and reaction kinetics, but will require very large amounts of highly accurate data for its inference, in order to avoid overfitting the model. This means that if a model has enough parameters, it can be fit very precisely to the dataset we initially collect, to the point that it cannot predict the outcome of a second perturbation that varies in some way. In contrast, a coarse-scale model with lower complexity may succeed in capturing high-level phenomena, such as which genes are ON or OFF, while requiring smaller amounts of more coarse-scale data. In course-scale models, we may miss many of the subtleties in states and fail to predict quite disparate outcomes resulting from branch points in the biological system under study.

In the context of genetic regulatory systems, fine-scale models, typically involving systems of differential equations, can be applied to relatively small and isolated genetic circuits for which many types of accurate measurements can be made. On the other hand, coarse-scale models are more suited to global (genome-wide) measurements, such as those produced with microarrays. Needless to say, according to the principle of Ockham's razor, which underlies all scientific theory building, model complexity should never be made higher than what is necessary to "explain the data" faithfully (Shmulevich et al. 2003).

4.7.2 How Do We Represent Networks?

The underlying assumption in creating a network model is that there is a collection of objects (such as genes or gene products) with at least pair-wise interactions connecting them. One way to conceptualize these interactions is to represent the components as nodes connected to each other by links or edges (most often directed edges), such that the edges represent the interactions between any 2 components. In this way, we can combine elements such as genes, proteins, metabolites, and other factors and represent them as a network, or graph, that can be modeled in a variety of ways.

Such a graph can be used to represent the relationships between elements in a network on 2 levels. First, the nature of the interactions between the elements can be reflected in the architecture of the graph (the patterns of nodes and edges) and the conditional relationships between the graph elements so that we know which elements respond to which other elements. Second, a set of network parameters can describe the strength of the dependencies between elements. For example, if 2 nodes, A and B, are connected by an edge, then A and B are dependent in some way. Such an interaction might model a transcription factor, B, that activates the expression of a particular gene encoding kinase, A. But if A and B are separated by a third node, C, then A and B are conditionally independent, given C. Here one might imagine that a transcription factor B activates expression of a second transcription factor C, which in turn induces expression of the kinase A gene.

To model biological systems, we have to define the rules by which B activates C and C activates A, taking into account important physical parameters, such as expression levels. We can, of course, build up even more complex interactions so that any node can have multiple incoming edges and, consequently, its activation can depend on complex interactions between those input edges. Further complexity is introduced when feedback loops are permitted in the model. To model biological systems, the challenge is to derive the network architecture and the set of rules by which the inputs to any node interact to produce the output.

A rather wide spectrum of approaches for modeling gene regulatory networks, each with its assumptions, data requirements, and goals, defines the nature and strength of the interactions. These include linear models (D'Haeseleer et al. 1999; van Someren et al. 2000), Bayesian networks (Murphy and Mian 1999; Friedman et al. 2000; Moler et al. 2000; Hartemink et al. 2001), neural networks (Weaver et al. 1999), differential equations (Mestl et al. 1995; Chen et al. 1999; Goutsias and Kim 2004), models including stochastic components on the molecular level (McAdams and Arkin 1997; Arkin et al. 1998). See Smolen et al. (2000), Hasty et al. (2001), and de Jong (2002) for reviews of general models.

The model system that has received, perhaps, the most attention is the Boolean network model originally introduced by Kauffman (Kauffman 1969; Glass and Kauffman 1973). Good reviews of this approach can be found in Kauffman (1993), Somogyi and Sniegoski (1996), Huang (1999), and Aldana et al. (2003). In this model, the state of a gene is represented by a Boolean variable (ON or OFF) and interactions between the genes are represented by Boolean logical functions, which determine the state of a gene on the basis of the states of some other genes. One of the appealing properties of Boolean networks is that they are inherently simple, emphasizing generic principles rather than quantitative biochemical details, but are able to capture the complex dynamics of gene regulatory networks. This is a valuable property for cross-species network inference and modeling because fine-scale experimental details are likely to differ between organisms. Computational models that reveal these logical interrelations have been successfully constructed from a variety of organisms such as *Drosophila*, sea urchin, *Arabidopsis thaliana*, and human (Bodnar 1997; Yuh et al. 1998; Mendoza et al. 1999; Huang and Ingber 2000).

4.7.3 To What Extent Do Such Models Represent Reality?

This is a question that can be asked of modeling in general. All models only approximate reality by means of some formal representation. In the context of Boolean networks as models of genetic regulatory networks, the binary approximation of gene expression (ON/OFF) is only suitable to capture aspects of regulation that possess a somewhat binary character. Although most biological phenomena are continuous rather than discrete, we often describe them in a binary logical language such as "on and off," "upregulated and downregulated," and "responsive and nonresponsive." Indeed, recent results suggest that gene regulation may often function digitally (Lahav et al. 2004). Before embarking on modeling gene regulatory networks using a Boolean formalism, it is prudent to test whether or not meaningful biological information can be extracted from gene expression data in which we make these simplifying assumptions.

This question was taken up by Shmulevich and Zhang (2002), who examined their ability to separate subclasses of tumors given the assumption that genes were expressed at only one of 2 levels, 0 or 1, within the data. This system simply asks whether subsets of genes can be identified that differentiate tumor subclasses, using only the detection of expression in the different samples rather than the level of expression. They reasoned that if the Boolean approximation failed in this relatively simple system, then there would be little hope for Boolean modeling of realistic genetic networks. Their results were promising.

By converting expression data from cDNA microarrays to this binary scale and using the Hamming distance to measure similarity, they distinguished different subtypes of gliomas (a similar experiment was also performed for sarcomas). Zhou et al. (2003b) used a similar approach in the context of classification. The revealing aspect of their work is that classification using binarized expressions proved to be only negligibly inferior to that using the original continuous expression values. The difference was that the genes derived via feature selection in the binary setting were different from the ones selected in the continuous. Together, these analyses suggest that a good deal of meaningful biological information is retained even when continuous expression levels are discretized and imply that Boolean approaches may be able to capture essential features of biological systems, at least in the context of predictive models.

4.7.4   Do We Have the "Right" Types of Data to Infer These Models?

All measurement techniques exhibit some degree of variability, including systematic and random noise, and functional genomics techniques are no exception. For cDNA microarray assays, the large number of genes surveyed and the sensitivity of the technique may in fact exacerbate the problem. Furthermore, genetic regulation exhibits considerable uncertainty on the biological level and some evidence even suggests that this type of noise is in fact advantageous in some regulatory mechanisms (McAdams and Arkin 1999). However, from a practical standpoint, the relatively limited amounts of data reflected in the small number of samples usually profiled in any experiment, combined with the experimental and biological noise that exists, can make quantitative inferences problematic and suggest that a coarse-scale approach may be most appropriate. Furthermore, most of the systems we would like to model are dynamic, but the vast majority of the data we have available are typically from a small number of fixed time points. Consequently, an important concern is the modeling of dynamic networks from steady-state data.

There are inherent limitations to this approach; steady-state behavior constrains the dynamical behavior, but does not determine it. Building a dynamical model from steady-state data is a form of overfitting. It is for this reason that we view a designed network as providing a regulatory structure consistent with the observed steady-state behavior. If our main interest is in steady-state behavior, then it is reasonable to try to understand dynamical regulation corresponding to steady-state behavior. Therefore, the answer to our question is that the data we have available constrains the class of the models we can construct and we have to choose our models appropriately.

4.7.5   Biological Systems Are Nonlinear Dynamical Systems

As noted previously, the transition from one organismal state to another is controlled by dynamic interactions, while biological homeostasis is mediated through complex nonlinear dynamical systems. Changes involving a few components can trigger a cascade of signals from the cell membrane all the way to the nuclear machinery to produce a global response. This ability to respond to the environment and its complex interactions, as well as to adapt through computing, is an important feature of complex biological systems. It is increasingly apparent that in order to understand the nature and context of cellular functions and the manner in which genes and molecules collectively form a biological system, it is necessary to study the behavior of genes and proteins in a holistic rather than an individual manner. Omics approaches allow us to collect the necessary data to begin to understand the underlying mechanisms. Thus, creating predictive models requires building models that capture the complexity of interactions, and it is useful to approach these problems from the general perspective of nonlinear dynamical systems (NDS).

Dynamical systems exhibit flows in a state space. A state of a system is information characterizing the situation at a given moment and the state space is the set of all possible states. It is important to note that there are a nearly infinite number of possible states; changing even one parameter defines a new state. However, natural constraints will narrow the set of unrestricted states to a subset of attainable states — that is, those states that actually have the possibility of occurring. It is essential, for example, in analyzing genetic networks to characterize the set of attainable states because these would shed light on the underlying biological mechanisms of genetic interactions.

Another important idea is that of an attractor — a region of the state space that captures the long-term behavior of the system (e.g., the collection of states in which an organism is able to live in a stable environment). A major discovery of the last few decades is that structurally stable systems can possess attractors on which the systems behave in an apparently random manner, a phenomenon referred to as chaotic dynamics. Dynamical systems theory provides a major source of understanding for the phenomenon of chaos. In fact, it is widely believed that complex and adaptable systems, such as the genome, operate in a zone between order and disorder, or on the "edge of chaos." As Stuart Kauffman (1995) puts it, "... a living system must first strike an internal compromise between malleability and stability. To survive in a variable environment, it must be stable to be sure, but not so stable that it remains forever static." Rule-based nonlinear dynamical system models, such as probabilistic Boolean networks (Shmulevich, Dougherty, Kim, et al. 2002), qualitatively reflect the nature of complex adaptive systems, in that they are "systems composed of interacting agents described in terms of rules" (Holland, 1995).

A central concept in NDS is that of structural stability, which represents the persistent behavior of a system under perturbation. Structural stability formally captures the idea of behavior that is not destroyed by small changes to the system. This is most certainly a property of real genetic networks because the cell must be able to maintain homeostasis in the face of external perturbations. Rule-based NDS models, such as Boolean networks, naturally capture this phenomenon as the system flows back into attainable states and attractors when some of its constituents (transcripts and proteins) are perturbed. In fact, Wolf and Eeckman (1998) showed in their study on the relationship between genomics regulatory element organization and gene regulatory dynamics that dynamic system behavior and stability of equilibria could be largely determined from regulatory element organization.

Perhaps equally important is the concept of uncertainty — in the model behavior as well as in the process of knowledge acquisition. The stochastic nature of genomics control, due to degradation of gene products, the spatial collision necessary before a reagent can exert its influence, and other causes, calls for probabilistic modeling methodologies. Thus, stochastic models are most likely to yield realistic results (McAdams and Arkin 1997). The field of information theory, traditionally used for communications technology applications, is well suited to study uncertainty measures, quantified through the use of probability theory.

Complex systems have the property that no single agent is singularly in control of the behavior of the system. Rather, control is dispersed among all agents, with varying levels of influence. Once again, this is the current view of genetic regulatory networks. In order to change the global behavior of the system significantly, it is necessary to act upon the key nodes that have a stronger influence over the system holistically. This fact is in tune with the inherent structural stability of genetic networks: The majority of small changes to the system are unlikely to alter its behavior globally. This issue can be properly addressed by the engineering discipline known as "control theory." One of the central problems in control theory is that of controllability: the problem of selecting an input so that the state of the system takes a desired value after some period of time. Many sophisticated optimization methods have been used for optimal control problems, including genetic algorithms (Krishnakumar and Goldberg 1992) and dynamic programming (Bertsekas 2001).

Numerous mathematical and computational methods have been proposed for construction of formal models of genetic interactions. Almost without exception, all these models essentially represent systems in that they characterize an interacting group of components forming a whole; they can also be viewed as a process that results in a transformation of signals and generates outputs in response to input stimuli. The models are dynamic in that they capture the time-varying quality of the physical process under study and can change their own behavior over time. Finally, the models can be considered to be generally nonlinear in that the interactions within the system yield behavior more complicated than the sum of the behaviors of the agents.

The preceding characteristics are representative of nonlinear dynamic systems. These are composed of states, input and output signals, transition operators between states, and output operators. In their most abstract form, they are very general. More mathematical structure is provided for particular application settings. For instance, in computer science they can be structured into the form of dataflow graphical networks that model asynchronous distributed computation — a model very close to genomics regulatory models. Based on long experience in electrical and computer engineering and more recent evidence from genomics, nonlinear dynamical systems appear to provide the appropriate framework to support the modeling of genomics systems. Indeed, most attempts to model gene regulatory networks fall within the scope of nonlinear dynamic systems.

In summary, nonlinear dynamic systems compose a proper framework for modeling and studying gene regulatory networks. The approach is highly multidisciplinary, requiring nonlinear signal processing, information theory, mathematical statistics, control theory, and computer science in order to study the dynamical behavior of these complex systems, develop stochastic models and tools to handle uncertainty, and develop efficient optimization algorithms to control the system behavior.

4.7.6 WHAT DO WE HOPE TO LEARN FROM THESE MODELS?

Our last question is concerned with what type of knowledge we hope to acquire with the chosen models and the available data. As a first step, we may be interested in discovering qualitative relationships underlying genetic regulation and control. That is, we wish to emphasize fundamental generic coarse-grained properties of large networks rather than quantitative details such as kinetic parameters of individual reactions (Huang 1999). Furthermore, we may wish to gain insight into the dynamical behavior of such networks and how they relate to underlying biological phenomena, such as cellular state dynamics. As an example, we may wish to predict the downstream effects of a targeted perturbation of a particular gene. Recent research indicates that many realistic biological questions may be answered within the seemingly simplistic Boolean formalism. Boolean networks are structurally simple, yet dynamically complex. They have yielded insights into the overall behavior of large genetic networks (Thomas et al. 1995; Somogyi and Sniegoski 1996; Szallasi and Liang 1998; Wuensche 1998) and allow the study of large datasets in a global fashion.

Perhaps the most salient limitation of standard Boolean networks is their inherent determinism. From a conceptual point of view, it is likely that the regularity of genetic function and interaction known to exist is not due to "hard-wired" logical rules, but rather to the intrinsic self-organizing stability of the dynamical system, despite the existence of stochastic components in the cell. From an empirical point of view, there are 2 immediate reasons why the assumption of only 1 logical rule per gene may lead to incorrect conclusions when inferring these rules from gene expression measurements: 1) The measurements are typically noisy and the number of samples is small relative to the number of parameters to be inferred, and 2) the measurements may be taken under different conditions and some rules may differ under these varying conditions.

During the past several years, a new mathematical rule-based model, called probabilistic Boolean networks (PBNs), has been developed to facilitate the construction of gene regulatory networks to assist scientists in revealing the intrinsic gene–gene relationships in cells and in exploring potential network-based strategies for therapeutic intervention (Kim et al. 2002; Shmulevich, Dougherty, Kim, et al. 2002; Shmulevich, Dougherty, and Zhang 2002a, 2002b, 2002c; Dougherty and Shmulevich 2003; Shmulevich et al. 2003; Datta et al. 2003, 2004; Zhou et al. 2003a; Hashimoto et al. 2004). There is already evidence that PBN models can reveal biologically relevant gene regulatory networks and can be used to predict the effects of targeted gene intervention. The problem with these nonlinear dynamic models is that they may be too complex for the data available and overfitting them may cause them to fail to accurately predict the outcome of a particular perturbation, such as the presence of a toxic compound.

What is important to remember is that network modeling cannot be conducted in isolation. Developing useful models relies on the close interplay between model development and experimental validation: The experiments drive the construction of models, which in turn suggest validation experiments. Only such a close interaction between wet lab experimental design and dry lab modeling and analysis will allow us to generate, refine, validate, and interpret the biologically relevant models.

4.8　THE PROBLEM OF VALIDATION

Functional genomics approaches provide global data on expression patterns that can be used in a range of applications. However, as is the case with any measurement, experimental noise and artifacts can contribute to inaccuracies. As such, it is imperative to confirm the results in some way. An additional potential problem with omics approaches is multiple testing. If, for example, we consider a microarray with 10,000 genes, applying a 95% confidence limit on gene selection ($p \leq .05$) means that, by chance, one would expect to find 500 genes as significant between the various phenotypic classes under study. While simply repeating the experiment, using a biological replicate, is sufficient to drop the statistical false positives to 25 (assuming independence), generally more stringent criteria than a simple p-value cut-off must be applied in selecting candidate genes.

Although many methods can be applied to adjust the probabilities derived from statistical tests, they often do little more than adjust the cut-offs used to select the significant genes. One critical test missing from simple computational approaches is the visual inspection of the complete expression profiles of the selected genes. Many confirmatory approaches involve follow-up on experiments focused on validating hypotheses from the large-scale analysis, while other techniques focus on confirming the observed levels in expression. In this latter area, we generally speak of 2 approaches: verification and validation.

Verification refers to the application of an independent analytical technique to assess the expression levels of a particular gene, protein, or metabolite in the same collection of samples. For example, one might apply qRT-PCR to assess the levels of gene expression observed in a particular collection of RNA samples using DNA microarrays. Verification experiments are important in that they can help to identify systematic biases introduced by a particular technology. However, verification experiments fail to address the more important issue of biological variability in samples.

Validation experiments, on the other hand, can address systematic and biological biases by using an independent set of test samples, generally, but not always, with an independent technique, and essentially validate the generality and extensibility of the result. In many ways, validation is a much stronger statement because it argues that the patterns of expression for particular genes that were observed can be generalized to other, independent biological samples. As noted previously, the choice of whether to verify or validate results and the methods used depend on the focus of the experiment, although, given the reliability of most commercial microarray platforms today, validation experiments are the most scientifically useful.

For mechanistic studies, validation is almost always essential and generally involves a first-pass validation of the expression levels observed, followed by detailed mechanistic studies to elucidate a hypothesized pathway or to demonstrate the central role a particular gene or gene product plays in producing some specific phenotype. Further evidence for the involvement of particular genes in response to a stimulus can come from previous knowledge (such as literature searches), functional analysis based on functional categories such as GO (the Gene Ontology Project; http://www.geneontology.org), assignment of genes to pathways, or other prior knowledge. Regardless of the weight of evidence supporting mechanistic hypothesis, direct testing is almost always necessary to validate the results. This is best done by making predictions based on the mechanistic hypothesis and then carrying out biological experiments to try to refute them.

For class discovery and classification experiments, the most important validation method is the application of the method to classification of data coming from an independent test set. Ideally, one should be able to assemble a classification training set of sufficient depth and breadth to identify appropriate signatures for prediction of phenotypic class, as well as a sufficiently large independent test set to allow the classification method to be tested and validated. In practice, microarray studies often have a limited number of samples and these are needed for building and training the algorithm. An alternative to using an independent test set is to do leave k out cross-validation (LKOCV). As one might guess, this approach leaves out k samples of the initial collection of N samples, develops a classifier using the $(N - k)$ samples that remain, and applies it to k samples in the test set. The classifier is judged on its ability to classify the k samples. This process is then repeated choosing a new set of k samples to be left out, the remaining $N - k$ used to create a new classifier, and this applied to the k samples left out, and so on. The properties of the classifiers are then weighted by their goodness in classifying the left-out samples and a single good classifier results. The simplest approach is simply to do leave one out cross-validation (LOOCV).

While this approach can be extremely useful when we lack an independent test set, it is often applied inappropriately as a partial rather than a full cross-validation. The distinction is the stage in the process where one "leaves k out." Many published studies have used their entire dataset to select a set of classification genes and then divide the samples into k and $(N - k)$ sample test and training sets. In fact, this has the potential to bias the results because the test and training sets are not independent because all of the samples were used to select the classification gene set. In particular, the presence of all of the samples in the initial gene selection process may favorably bias the ultimate success of any classifier that is constructed.

In full LKOCV, the data are divided into k and $(N - k)$ sample test and training sets and the $(N - k)$ training set is used to select a classification gene set and then to apply it to creating a classification algorithm and using it to classify the k test samples. One can then estimate the accuracy of the classification system by simply averaging over the complete set of classifiers.

Developing a biological interpretation for any classification experiment can be difficult because the genes providing the best discriminating power between classes may very often be difficult to link casually or mechanistically to the underlying disease. LKOCV approaches present the further challenge that each iteration may produce a unique set of classification genes with a very small number in common to all of these (although further insight might come from using genes appearing in a majority of iterations or by taking the union of all iterations). Analyzing these gene lists generally follows the same approach we use in a discovery experiment and the most effective approaches come from linking these gene lists to ancillary information such as polymorphism data or functional roles.

Regardless, it may be that the classification gene set cannot be linked to the disease given current knowledge. In this context, it is useful to note that many examples of biomarkers of unknown function in the context of the disease are extremely useful as diagnostic or prognostic markers for various diseases. In some sense, we can consider the gene lists emerging from classification experiments as sets of biomarkers that have potentially important applications; if these have some biological interpretation, this is a significant bonus that can shed light on the underlying biological mechanism.

4.9 PREDICTIVE TOXICOLOGY

Predictive toxicology refers to the science of making prospective predictions of toxicity outcomes based on previously untested relationships. One of the aims of toxicology is to predict the effects of various compounds at the organism and population levels. Many strategies have been developed in the ecology and toxicology research communities for monitoring the physiological endpoints of various carcinogens, neurotoxicants, hepatotoxicants, etc. While multiple approaches can be used to accomplish this goal, many believe that toxicogenomics and the new omics era offer great promise in supporting the development of this field. Functional genomics technologies can complement the repertoire of assays available for understanding the mechanisms and mode of action of such compounds. For example, rather than having a single sensor to measure the LC50s for compound X, we now have access to thousands of sensors to monitor how the molecular state of a cell is perturbed in response to compound X to cause toxicity.

Predicting toxic effects will be challenging because multiple mechanisms of toxic action and species-specific responses usually exist. Prediction of toxicity outcomes will require the combination of chemical descriptors, statistical methodology, and acquisition of large datasets to link a given descriptor to a specific response. One possibility now available — the focus of this section — is the ability to monitor genetic pathways conserved across a wide range of life forms. Genes whose products operate together to bring about a specific function are known as a module. Core modules are modules conserved across species. Different core modules can be defined at different nodes in the phylogenetic tree that relates a set of organisms. For example, a cell-cycle module is ancient and logically resides at the root of the tree, while apoptosis is younger (more generally useful in multicellular organisms) and logically resides at the ancestral node relating all multicellular animals.

The pattern of activation and deactivation of core modules reports on the state of a cell, which in turn reports on the vigor of a tissue or organ system, which in turn reports on the health of the organism or population. It seems promising, therefore, to monitor the state of these molecular core modules to be able to link phenotypic outcomes across multiple species. It is of fundamental interest to monitor and describe the patterns of activation and deactivation of core modules and to discern how they discriminate chemical reactivities and toxicological outcomes across species.

Comparative genomics research leads to the identification and functional classification of orthologous genes, pathways, and networks across multiple species. By studying the coordinate expression of genes and proteins through molecular profiling, it is possible to glean information on the physiological status of genes and gene interactions in cellular and developmental processes and in disease. Such studies have succeeded in defining the core biological functional relationships of genes, pathways, and networks in ecologically relevant organisms as well as in model laboratory organisms. In addition, there are large conventional toxicology and ecotoxicology databases that serve to document the phenotypes that result from environmental stressors and toxicants.

Toxicogenomics research is now defining how the genomes of these same organisms are involved in responses to the same agents. However, the number of toxicants and stressors evaluated to date is too small to provide the database for predictive toxicology based on the transcriptions, proteomics, or metabolomics responses. Thus, a strategy must be devised to leverage the existing toxicology and ecotoxicology databases in defining adverse outcomes, and this strategy must certainly take into account the knowledge that has evolved relating chemical structure, inherent chemical–biological reactivity, and toxicological effects.

There are likely fewer than 30 categories of chemical–biological reactivity, including specific and nonspecific reactivities that result in, for example, uncoupling of oxidative phosphorylation, DNA and protein alkylations, chemical–receptor noncovalent interactions, hormonal mediators, etc. Fortuitously, there are also fewer than 30 categories of standard toxicological endpoints. These endpoints include acute general or target organ toxicity, developmental toxicity, reproductive toxicity, immunotoxicity, hepatotoxicity, cardiovascular toxicity, nephrotoxicity, genotoxicity, carcinogenicity, etc. in humans and ecological species of concern. Endpoints such as growth, survival, and reproduction success are especially useful as ecological outcomes in environmental protection regulation. Similarly, carcinogenicity and developmental and reproductive toxicity are key regulatory endpoints of concern for health risk assessment.

It is already clear from the limited toxicogenomics investigations that have been performed that toxicogenomics is sufficiently robust to demonstrate signatures of chemical exposure and effect and that these signatures can be correlated with phenotypes that correspond to conventional toxicological endpoints. Thus, it has been demonstrated that these exposure- and outcome-specific patterns of altered gene, protein, and endogenous metabolite profiles can provide insights into modes or mechanisms of toxicant action (Waring et al. 2001; Aardema and MacGregor 2002; Hamadeh, Bushel, Jayadev, DiSorbo, et al. 2002; Hamadeh, Bushel, Jayadev, Martin, et al. 2002; Johnson et al. 2003, 2004; Toraason et al. 2004), disease causation (Lu et al. 2001; Iida et al. 2003; Kramer, Curtiss et al. 2004), and human susceptibility (Blair et al. 2002; Balmain et al. 2003; Guerreiro et al. 2003; Weiss et al. 2004).

Therefore, taking into account the available tools of genomics and the results of toxicogenomics studies to date as well as the large toxicological databases that serve now to provide the bases for predictive toxicology, quantitative structure activity analysis, and chemometrics, it is feasible to define a strategy for the development of a new predictive toxicology paradigm based on conventional toxicology, functional genomics, and toxicogenomics. In outline format, the strategy would be carried out as follows:

1) Use comparative functional genomics to identify evolutionarily conserved genes, promoter elements, pathways, and networks that reflect basic or core biological modules across species under conditions of toxicity and to evaluate toxic effects.
2) Evaluate the panel of chemicals representing reactivities for their effects on core modules and their predictive power to discriminate toxicological endpoints.
3) Identify known outcomes within existing toxicology and ecotoxicological databases that can validate the predictive capability of this approach.
4) Add additional chemical species that represent the range of members of a chemical reactivity classification. Determine the predictive value of toxicogenomics analysis as compared to conventional structure activity analysis and toxicological evaluation (Figure 4.6).

Molecular profiling of gene, protein, metabolite datasets is used to define core biological processes. Starting with a representative organism (training species) within a phylogenetic node, exhaustive chemical screening and expression profiling are conducted to generate Module profiles indicative of toxicity endpoints. Subsequently, a test organism is challenged with a chemical stressor to generate expression data. These data are segregated into core modules that are predictive of toxicological endpoints. Subsequent validation studies can be conducted to confirm biological effects.

4.10 EDUCATING THE COMMUNITY

It is widely recognized that tackling important and challenging biomedical problems requires augmentation and complementation of the research methods in biology with the approaches and methods of nonbiological scientific disciplines. It is essential that biologists work closely with specialists in other fields who have extensive experience in looking at the same types of problems in their own domains; nevertheless, there remains a critical shortage of such scientists. In a recent editorial, "The Human Genome Project: Help Wanted" in *IEEE Spectrum*, the flagship publication of the world's largest technical professional association, it was stated that "…getting to this next level of understanding will require engineers and computer scientists and mathematicians and physicists who can work with genomics issues, so that they can not only support the biological enterprise but also help drive it. Their task will not just be one of engineering devices and programs, but one of engineering ideas." At the same time, it is critical that molecular biologists be capable of taking multidisciplinary approaches to address important biological research problems.

To achieve these goals, cross-disciplinary training is a key element for enhancing an environment that will facilitate educating future researchers in computational and

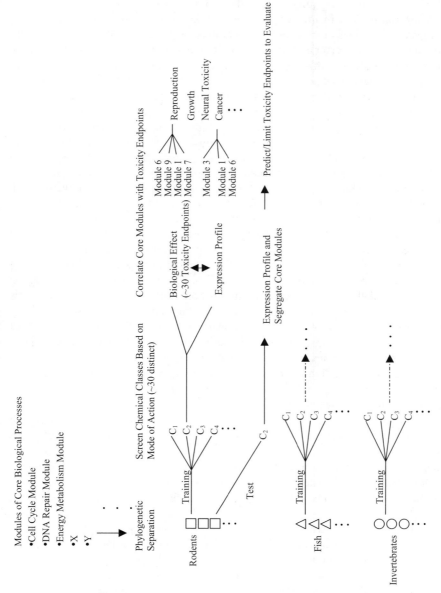

FIGURE 4.6 Phylogenetically informed extrapolations to cross-species predictions of toxicology based on core biological processes.

systems biology. New training programs need to foster a new, fundamental scientific discipline that will bring to genomics, proteomics, and metabolomics the kind of structural, model-based analysis and synthesis that forms the basis of mathematically rigorous engineering. In such programs, it is crucial to emphasize the strong mathematical underpinnings required for the development of new, complex models in the omics disciplines and to involve students in functional genomics courses with an eye towards the discovery of underlying universal principles. It is important that graduates possess a greater ability to internalize the mathematics–biology integration than those who come to this field later in their careers. Needless to say, students in such programs must have the benefit of a strong multidisciplinary faculty who should design a core curriculum that includes biological, engineering, mathematical, statistical, physical, and computer sciences.

The early stages of developing such multidisciplinary programs present their own challenges. For example, core courses must be organized in such a way that they are tailored to students with different backgrounds. Students with engineering backgrounds should enroll in introductory courses in genomics, while biology students should participate in courses that teach engineering principles. However, such introductory courses must be quite different from the traditional introductory courses given in biology and engineering departments because they should assume some level of background knowledge in the student's area of study. For instance, a student with a B.S. degree in biology who takes an engineering methods course should have the opportunity to focus immediately on the engineering aspects without spending too much time on the biological preliminaries. Similarly, a course focusing on probability, statistics, and pattern recognition should not be a generic course that is commonly taught in engineering departments, but rather should be specifically tailored to biological problems and applications. Finally, in the later stages of multidisciplinary programs, students should be encouraged to enroll in advanced courses in disciplines relevant to the students' research work.

Research and education are tightly intertwined. Cross-disciplinary integration is equally important for research. An effective way to achieve this goal is to have an integrated program in which biologists and modeling scientists work together with constant communication and feedback and with immediate and spontaneous dry lab and wet lab interactions in a research mode. This is one of the most difficult tasks facing integrative biology and toxicology research. Interactions and collaborations among modeling scientists or among biologists have been ongoing for years and are supported by various funding mechanisms. However, the integration between biologists and modeling scientists is not a norm yet and faces the following obstacles:

- These scientists are usually located in physically separated departments and institutions and usually speak different professional languages. There is no environment for daily communication and mutual learning.
- There are very few funding mechanisms to support integration across different disciplines, though this is slowly starting to change.
- Long-term institutional support is often hinged on the potential for receiving extramural funding. With the previously listed obstacles, most institutions have difficulty in providing sustained support to the integration effort.

Because of these obstacles, such integration has been slow and limited, deeply frustrating researchers making sincere efforts to bridge disciplines in order to push biology and toxicology research onto the path of systems biology.

4.11 RECOMMENDATIONS FOR ADVANCING THE FIELD

Realization of the promise of toxicogenomics technologies in predictive human and environmental risk assessment will require the use of well-defined experimental systems and animal models to develop richer databases that can be used to elucidate mechanisms of toxic action, identify potentially harmful substances, and predict toxicity outcomes. To achieve these goals, it will be necessary to focus on conserved core biological processes for development of

- experimental models that extend beyond the laboratory field and represent conserved phylogenetic relationships,
- new algorithms and methodologies that help uncover core biological processes,
- standardized orthology resources,
- new methodologies for deconvolution of products arising from different portions of a gene,
- tools that optimize database alignment on a per-probe basis and that facilitate database updates,
- centralized public toxicogenomics databases that can be readily updated and are expandable,
- universal standards for proteomics and metabolomics databases,
- network modeling tools that assimilate phylogenetic similarities and differences to identify subnetworks that represent common as well as individual properties, and
- resource inventories that include protein–protein interactions and micro-RNAs.

4.12 CONCLUDING REMARKS

During the past few years there have been many discussions in the literature on noise in microarray assays, disparate results arising from the use of different platforms, questions regarding the validity of results, and the need to validate the findings. If one closely examines the underlying issues, it is clear that functional genomics assays are no different from any other approach to assaying levels of gene expression: Each has its biases and limitations. Microarrays simply provide much more data than do techniques such as qRT-PCR or northern blots and the likelihood of false positives and false negatives increases as the number of genes assayed increases — a manifestation of the multiple testing problem. Underlying all of these issues are efforts to understand what can be done with the data that emerge. Although there are no absolute answers, some overarching generalizations can be made that will help guide the follow-on experiments.

First, whether one is working on a mechanistic study or trying to identify genes that can be used for sample classification, microarray assays generally provide lists of genes that can be significantly correlated with some classes and, therefore, should be treated not as truths, but as hypotheses that can be tested.

Second, statistical significance is fine, but biological significance is better. Statistics provides powerful tools for identifying candidate genes, for prioritizing them in the lack of any other evidence, and in helping to resolve features in the data. But if 30 of the top 50 genes of interest are energy metabolism genes, the working hypothesis is likely to be that the experimental system involves changes in energy metabolism regardless of whether these genes are the first or the last 30 on the gene list.

Third, it is important to note that any change in how the analysis is completed is likely to change what is identified as significant. Starting with laboratory protocols and ending with parameters for slide scanning, image processing, data normalization, and the choice of the analysis tools, all these parameters contribute significantly to outcome. One way to approach this problem is to apply multiple approaches and then look for a common set of "high-confidence" genes. Another way that may be more useful is to look at pathways and functional classes in order to identify common biological themes that overlie all of the analyses.

Fourth, in confirming any microarray result it is often useful to use an alternate technique to assay gene expression levels. Finally, microarrays only provide correlations between a particular pattern of expression and some biological class. The real biology is not on the array, but on the bench back in the laboratory. Microarray experiments can be powerful tools for developing testable hypotheses and can even play a significant role in conducting such tests. The value of functional genomics experiments is that they provide unbiased surveys of large numbers of genes, and this can be extremely powerful for discovering potential new mechanisms and new subgroups within classes. But, ultimately, microarrays remain a tool for discovery and bioinformatics is simply a filter that can increase the power of the array experiment.

REFERENCES

Aardema MJ, MacGregor JT. 2002. Toxicology and genetic toxicology in the new era of "toxicogenomics": impact of "-omics" technologies. Mutat Res 499:13–25.

Aldana M, Coppersmith S, Kadanoff LP. 2003. Boolean dynamics with random couplings. In: Kaplan E, Marsden JE, Sreenivasan KR, editors. Perspectives and problems in nonlinear science. Springer Applied Mathematical Sciences Series, Dordrecht: Springer-Verlag.

Alon U, Barkai N, Notterman DA, Gish K, Ybarra S, Mack D, Levine AJ. 1999. Broad patterns of gene expression revealed by clustering analysis of tumor and normal colon tissues probed by oligonucleotide arrays. Proc Natl Acad Sci U S A 96:6745–6750.

Altschul SF, Gish W, Miller W, Myers RW, Lipman DJ. 1990. Basic local alignment search tool. J Mol Biol 215:403–410.

Amato U, Larobina M, Antoniades A, Alfano B. 2003. Segmentation of magnetic resonance brain images through discriminant analysis. J Neuroscience Meth 131:65–74.

Arkin A, Ross J, McAdams HH. 1998. Stochastic kinetic analysis of developmental pathway bifurcation in phage-infected *Escherichia coli* cells. Genetics 149:1633–1648.

Ball CA, Sherlock G, Parkinson H, Rocca-Sera P, Brooksbank C, Causton HC, Cavalieri D, Gaasterland T, Hingamp P, Holstege F, et al. 2002. Standards for microarray data. Science 298(5593):539.

Ball G, Mian S, Holding F, Allibone RO, Lowe J, Ali S, Li G, McCardle S, Ellis O, Greaser C, Rees RC. 2002. An integrated approach utilizing artificial neural networks and SELDI mass spectrometry for the classification of human tumours and rapid identification of potential biomarkers. Bioinformatics 18:395–404.

Balmain A, Gray J, Ponder B. 2003. The genetics and genomics of cancer. Nat Genet 33(Suppl):234–244.

Begley TJ, Rosenbach AS, Ideker T, Samson LD. 2002. Damage recovery pathways in *Saccharomyces cerevisiae* revealed by genomic phenotyping and interactome mapping. Mol Cancer Res 1:103–112.

Bertsekas D. 2001. Dynamic programming and optimal control. 2nd ed. Nashua, NH: Athena Scientific.

Blair IP, Adams LJ, Badenhop RF, Moses MJ, Scimone A, Morris JA, Ma L, Austin CP, Donald JA, Mitchell PB, and others. 2002. A transcript map encompassing a susceptibility locus for bipolar affective disorder on chromosome 4q35. Mol Psychiatry 7(8):867–873.

Bodnar JW. 1997. Programming the Drosophila embryo. J Theor Biol 188:391–445.

Boulesteix A-L, Tutz G, Stimmer K. 2003. A cart-based approach to discover emerging patterns in microarray data. Bioinformatics 19:2465–2472.

Boorman GA, Anderson SP, Casey WM, Brown RH, Crosby LM, Gottschalk K, Easton M, Ni H, Morgan KT. 2002. Toxicogenomics, drug discovery, and the pathologist. Toxicol Pathol 30(1):15–27.

Brazma A, Sarkans U, Robinson A, Vilo J, Vingron M, Hoheisel J, Fellenberg K. 2002 Microarray data representation, annotation and storage. Adv Biochem Eng Biotechnol 77:113–139.

Brazma A, Hingamp P, Quackenbush J, Sherlock G, Spellman P, Stoeckert C, Aach J, Ansorge W, Ball CA, Causton HC, and others. 2001. Minimum information about a microarray experiment (MIAME)-toward standards for microarray data. Nat Genet 29:365–371.

Brazma A, Parkinson H, Sarkans U, Shojatalab M, Vilo J, Abeygunawardena N, Holloway E, Kapushesky M, Kemmeren P, Lara GG, et al. 2003. ArrayExpress—a public repository for microarray gene expression data at the EBI. Nucleic Acids Res 31:68–71.

Bumm K, Zheng M, Bailey C, Zhan F, Chiriva-Internati M, Eddlemon P, Terry J, Barlogie B, Shaughnessy JD Jr. 2002. CGO: utilizing and integrating gene expression microarray data in clinical research and data management. Bioinformatics 18:327–328.

Bushel PR, Hamadeh H, Bennett L, Sieber S, Martin K, Nuwaysir EF, Johnson K, Reynolds K, Paules RS, Afshari CA. 2001. MAPS: a microarray project system for gene expression experiment information and data validation. Bioinformatics 17:564–565.

Chaussabel D, Sher A. 2002. Mining microarray expression data by literature profiling. Genome Biol. E pub 2002 Sept 13;3(10):RESEARCH0055.

Chen T, He HL, Church GM. 1999. Modeling gene expression with differential equations. Pacific Symp Biocomp 4:29–40.

Chu TM, Deng S, Wolfinger R, Paules RS, Hamadeh HK. 2004. Cross-site comparison of gene expression data reveals high similarity. Environ Health Perspect 112:449–455.

Coen M, Ruepp SU, Lindon JC, Nicholson JK, Pognan F, Lenz EM, Wilson ID. 2004. Integrated application of transcriptomics and metabonomics yields new insight into the toxicity due to paracetamol in the mouse. J Pharm Biomed Anal 35:93–105.

Cutler P, Bell DJ, Birrell HC, Connelly JC, Connor SC, Holmes E, Mitchell BC, Monte SY, Neville BA, Pickford R, et al. 1998. An integrated proteomic approach to studying glomerular nephrotoxicity. Electrophoresis 20:3647–3658.

Datta A, Choudhary A, Bittner ML, Dougherty ER. 2003. External control in Markovian genetic regulatory networks. Machine Learning 52:169–181.

Datta A, Choudhary A, Bittner ML, Dougherty ER. 2004. External control in Markovian genetic regulatory networks: The imperfect information case. Bioinformatics 20(6):924–930.

de Jong H. 2002. Modeling and simulation of genetic regulatory systems: a literature review. J Comp Biol 9(1):69–103.

DeRisi J, Penland L, Brown PO, Bittner ML, Meltzer PS, Ray M, Chen Y, Su YA, Trent JM. 1996. Use of a cDNA micro array to analyse gene expression patterns in human cancer. Nat Genet 14:457–460.

D'Haeseleer P, Wen X, Fuhrman S, Somogyi R. 1999. Linear modeling of mRNA expression levels during CNS development and injury. Pacific Symp Biocomp 4:41–52.

Dougherty ER, Shmulevich I. 2003. Mappings between probabilistic boolean networks. Signal Proc 83(4):799–809.

Eckman BA, Kosky AS, Laroco LA Jr. 2001. Extending traditional query-based integration approaches for functional characterization of post-genomic data. Bioinformatics 17:587–601.

Edgar R, Domrachev M, Lash AE. 2002.. Gene expression omnibus: NCBI gene expression and hybridization array data repository. Nucleic Acids Res 30:207–210.

Eisen MB, Spellman PT, Brown PO, Botstein D. 1998. Cluster analysis and display of genome-wide expression patterns. Proc Natl Acad Sci U S A 95:14863–14868.

Ermolaeva O, Rastogi M, Pruitt KD, Schuler GD, Bittner ML, Chen Y, Simon R, Meltzer P, Trent JM, Boguski MS. 1998. Data management and analysis for gene expression arrays. Nat Genet 20:19–23.

Fountoulakis M, Berndt P, Boelsterli UA, Crameri F, Winter M, Albertini S, Suter L. 2000. Two-dimensional database of mouse liver proteins: changes in hepatic protein levels following treatment with acetaminophen or its nontoxic regioisomer 3-acetamido-phenol. Electrophoresis 21:2148–2161.

Friedman N, Linial M, Nachman I, Pe'er D. 2000. Using bayesian network to analyze expression data. J Comput Biol 7:601–620.

Furey TS, Cristianini N, Duffy N, Bednarski DW, Schummer M, Haussler D. 2000. Support vector machine classification and validation of cancer tissue samples using microarray expression data.Bioinformatics 16(10):906–914.

Glass L, Kauffman SA. 1973. The logical analysis of continuous, non-linear biochemical control networks. Journal of Theor Biol 39:103–129.

Golub TR, Slonim DK, Tamayo P, Huard C, Gaasenbeek M, Mesirov JP, Coller H, Loh ML, Downing JR, Caligiuri MA, et al. 1999. Molecular classification of cancer: class discovery and class prediction by gene expression monitoring. Science 286(5439):531–537.

Goutsias J, Kim S. 2004. A nonlinear discrete dynamical model for transcriptional regulation: construction and properties. Biophys J 86:1922–1945.

Guerreiro N, Staedtler F, Grenet 0, Kehren J, Chibout SD. 2003. Toxicogenomics in drug development. Toxicol Pathol 31:471–479.

Hamadeh HK, Bushel PR, Jayadev S, Martin K, DiSorbo O, Sieber S, Bennett L, Tennant R, Stoll R, Barrett JC, et al. 2002a. Gene expression analysis reveals chemical-specific profiles. Toxicol Sci 67:219–231.

Hamadeh HK, Bushel PR, Jayadev S, DiSorbo O, Bennett L, Li L, Tennant R, Stoll R, Barrett JC, Paules RS, et al. 2002b Prediction of compound signature using high density gene expression profiling. Toxicol Sci 67:232–240.

Hamadeh HK, Jayadev S, Gaillard ET, Huang Q, Stoll R, Blanchard K, Chou J, Tucker CJ, Collins J, Maronpot R, and others. 2004. Integration of clinical and gene expression endpoints to explore furan-mediated hepatotoxicity. Mutat Res 549:169–183.

Hartemink AJ, Gifford DK, Jaakkola TS, Young RA. 2001. Using graphical models and genomic expression data to statistically validate models of genetic regulatory networks. Pacific Symposium on Biocomputing: January 2001; Hawaii.

Hashimoto RF, Kim S, Shmulevich I, Zhang W, Bittner ML, Dougherty ER. 2004. Growing genetic regulatory networks from seed genes. Bioinformatics 20(8):1241–1247.

Hasty J, McMillen D, Isaacs F, Collins JJ. 2001. Computational studies of gene regulatory networks: *in numero* molecular biology. Nat Rev Genet 2:268–279.

Hermjakob HI, Montecchi-Palazzi L, Bader G, Wojcik J, Salwinski L, Ceol A, Moore S, Orchard S, Sarkans U, von Mering C, et al. 2004. The HUPO PSI's molecular interaction format—a community standard for the representation of protein interaction data. Nat Biotechnol 22:177–183.

Hogstrand C, Balesaria S, Glover CN. 2002. Application of genomics and proteomics for study of the integrated response to zinc exposure in a non-model fish species, the rainbow trout. Comp Biochem Physiol Part B Biochem Mol Biol 133:523–535.

Holland JH. 1995. Hidden order: how adaptation builds complexity, Reading, MA: Helix Books.

Huang S. 1999. Gene expression profiling, genetic networks, and cellular states: an integrating concept for tumorigenesis and drug discovery. J Molec Med 77:469–480.

Huang S, Ingber DE. 2000. Regulation of cell cycle and gene activity patterns by cell shape: evidence for attractors in real regulatory networks and the selective mode of cellular control. Int J Genet 238. http://www.interjournal.org.

Hughes TR, Marton MJ, Jones AR, Roberts CJ, Stoughton R, Armour CD, Bennett HA, Coffey E, Dai H, He YD, and others. 2000. Functional discovery via a compendium of expression profiles. Cell 102:109–126.

Ideker T, Galitski T, Hood L. 2001. A new approach to decoding life: systems biology. Annu Rev Genomics Hum Genet 2:343–372.

Ihmels J, Bergmann S, Barkai N. 2004. Defining transcription modules using large-scale gene expression data. Bioinformatics 20(13):1993–2003. Epub 2004 Mar 25.

Iida M, Anna CH, Hartis J, Bruno M, Wetmore B, Dubin JR, Sieber S, Bennett L, Cunningham ML, Paules RS, et al. 2003. Changes in global gene and protein expression during early mouse liver carcinogenesis induced by non-genotoxic model carcinogens oxazepam and Wyeth-14, 643. Carcinogenesis 24:757–770.

Jiang M, Ryu J, Kirally M, Duke K, Reinke V, Kim SK. 2001. Genome-wide analysis of developmental and sex-regulated gene expression profiles in *Caenorhabditis elegans*. Proc Nat Acad Sci U S A 98:219–223.

Jiang XS, Zhou H, Zhang L, Sheng QH, Li SJ, Li L, Hao P, Li YX, Xia QC, Wu JR, et al. 2004. A high-throughput approach for subcellular proteome: identification of rat liver proteins using subcellular fractionation coupled with two-dimensional liquid chromatography tandem mass spectrometry and bioinformatic analysis. Mol Cell Proteomics 3:441–455.

Johnson, CD, Balagurunathan Y, Lu KP, Tadesse M, Falahatpisheh MH, Carroll RJ, Dougherty ER, Afshari CA, Ramos KS. 2003. Genomic profiles and predictive biological networks in oxidant-induced atherogenesis. Physiol Genom 13:263–75.

Johnson CD, Balagurunathan Y, Tadesse MG, Falahatpisheh MH, Walker MK, Dougherty ER, Ramos KS. 2004. Unraveling gene-gene interactions regulated by ligands of the aryl hydrocarbon receptor. Environ Health Persp 112:403–412.

Jones A, Hunt E, Wastling J, Pizarro A, Stoeckert CJ Jr. 2004. An object model and database for functional genomics. Bioinformatics 20(10)1583-1590. Epub 2004 May 14.

Kauffman SA. 1969. Metabolic stability and epigenesis in randomly constructed genetic nets. J Theor Biol 22:437–467.

Kauffman SA. 1993. The origins of order: Self-organization and selection in evolution. New York: Oxford University Press.

Kauffman SA. 1995. At home in the universe. New York: Oxford University Press.

Kim S, Dougherty ER, Chen Y, Sivakumar K, Meltzer P, Trent JM, Bittner M. 2000a. Multivariate measurement of gene expression relationships. Genomics 67:201–209.

Kita Y, Shiozawa M, Jin W, Majewski RR, Besharse JC, Greene AS, Jacob HJ. 2002. Implications of circadian gene expression in kidney, liver and the effects of fasting on pharmacogenomic studies. Pharmacogenetics 12:55–65.

Kramer JA, Curtiss SW, Kolaja KL, Alden CL, Blomme EAG, Curtiss WC, Davila JC, Jackson CJ, Bunch RT. 2004. Acute molecular markers of rodent hepatic carcinogenesis identified by transcription profiling. Chem Res Toxicology 17:463–470.

Kramer JA, Pettit SD, Amin RP, Bertram TA, Car B, Cunningham M, Curtiss SW, Davis JW, Kind C, Lawton M, and others. 2004. Overview on the application of transcription profiling using selected nephrotoxicants for toxicology assessment. Environ Health Perspect 112:460–464.

Krishnakumar K, Goldberg DE. 1992. Control system optimization using genetic algorithms. J Guidance Control Dyn 15(3):735–740.

Lahav G, Rosenfeld N, Sigal A, Geva-Zatorsky N, Levine AJ, Elowitz MB, Alou U. 2004. Dynamics of the p53-Mdm2 feedback loop in individual cells. Nat Gen 36:147–150.

Lee JS, Chu IS, Mikaelyan A, Calvisi DF, Heo J, Reddy JK, Thorgeirsson SS. 2004. Application of comparative functional genomics to identify best-fit mouse models to study human cancer. Nat Genet 36:1306–1311.

Liao B, Hale W, Epstein CB, Butow RA, Garner HR. 2000. MAD: a suite of tools for microarray data management and processing. Bioinformatics 16:946–947.

Lipshutz RJ, Morris D, Chee M, Hubbell E, Kozal MJ, Shah N, Shen N, Yang R, Fodor SP. 1995. Using oligonucleotide probe arrays to access genetic diversity. Biotechniques 19(3):442–447.

Lu T, Liu J, LeCluyse EL, Zhou YS, Cheng ML, Waalkes MP. 2001. Application of cDNA microarray to the study of arsenic-induced liver diseases in the population of Guizhou, China. Toxicol Sci 59:185–192.

Mattes WB, Pettit SD, Sansone SA, Bushel PR, Waters MD. 2004. Database development in toxicogenomics: issues and efforts. Environ Health Perspect 112:495–505.

McAdams HH, Arkin A. 1997. Stochastic mechanisms in gene expression. Proc Natl Acad Sci USA 94:814–819.

McAdams HH, Arkin A. 1999. It's a noisy business! Genetic regulation at the nanomolar scale. Trends Genet 15:65–69.

Mendoza L, Thieffry D, Alvarez-Buylla ER. 1999. Genetic control of flower morphogenesis in Arabidopsis thaliana: a logical analysis. Bioinformatics 15:593–606.

Mestl T, Plahte E, Omholt SW. 1995. A mathematical framework for describing and analysing gene regulatory networks. J Theoretical Biol 176(2):291–300.

Moch H, Schraml P, Bubendorf L, Mirlacher M, Kononen J. Gasser T, Mihatsch MJ, Kallioniemi OP, Sauter G. 1999. High-throughput tissue microarray analysis to evaluate genes uncovered by cDNA microarray screening in renal cell carcinoma. Am J Pathol 154(4):981–986.

Moler EJ, Radisky DC, Mian IS. 2000. Integrating naive Bayes models and external knowledge to examine copper and iron homeostasis is S. cerevisiae. Physiol Genom 4:127–135.

Murphy K, Mian S. 1999. Modelling gene expression data using dynamic bayesian networks. Technical Report, University of California, Berkeley, CA.

Newton RK, Aardema M, Aubrecht J. 2004. The utility of DNA microarrays for characterizing genotoxicity. Environ Health Perspect 112:420–422.

Nurse P. 2003. Understanding cells. Nat Biotechnol 424:883.

Orchard S, Hermjakob H, Julian RK Jr, Runte K, Sherman D, Wojcik J, Zhu W, Apweiler R. 2004. Common interchange standards for proteomics data: Public availability of tools and schema. Proteomics 4:490–491.

Perou CM Jeffrey SS, van de Rijn M, Rees CA, Eisen MB, Ross DT, Pergamenschikov A, Williams CF, Zhu SX, Lee JC, et al. 1999. Distinctive gene expression patterns in human mammary epithelial cells and breast cancers. Proc Natl Acad Sci USA 96:9212–9217.

Petricoin EF, Hackett JL, Lesko LJ, Puri RK, Gutman SI, Chumakov K, Woodcock J, Feigal DW Jr, Zoon KC, Sistare FD. 2002. Medical applications of microarray technologies: a regulatory science perspective. Nat Genet 32(Suppl):474–479.

Ramaswamy S, Tamayo P, Rifkin R, Mukherjee S, Yeang CH, Angelo M, Ladd C, Reich M, Latulippe E, Mesirov JP, et al. 2001. Multiclass cancer diagnosis using tumor gene expression signatures. Proc Natl Acad Sci U S A 98(26):15149–15154. Epub 2001 Dec 11.

Ramos KS. 2003. EHP toxicogenomics: a publication forum in the postgenome era. EHP Toxicogenom 111(1T):A13.

Ruepp SU, Tonge RP, Shaw J, Wallis N, Pognan F. 2002. Genomics and proteomics analysis of acetaminophen toxicity in mouse liver. Toxicol Sci 65:135–150.

Shmulevich I. 2003. Model selection in genomics. Environ Health Persp Toxicogenom 111(6):A328–A329.

Shmulevich I, Dougherty ER, Kim S, Zhang W. 2002a. Probabilistic boolean networks: a rule-based uncertainty model for gene regulatory networks. Bioinformatics 18(2):261–274.

Shmulevich I, Dougherty ER, Zhang W. 2002d. Gene perturbation and intervention in probabilistic boolean networks. Bioinformatics 18(10):1319–1331.

Shmulevich I, Dougherty ER, Zhang W. 2002e. Control of stationary behavior in probabilistic boolean networks by means of structural intervention. J Biol Sys 10(4):431–445.

Shmulevich I, Dougherty ER, Zhang W. 2002f. From boolean to probabilistic boolean networks as models of genetic regulatory networks. Proc IEEE 90(11):1778–1792.

Shmulevich I, Gluhovsky I, Hashimoto R, Dougherty ER, Zhang W. 2003a. Steady-state analysis of probabilistic boolean networks. Comp Funct Genom 4(6):601–608.

Shmulevich I, Zhang W. 2002. Binary analysis and optimization-based normalization of gene expression data. Bioinformatics 18(4):555–565.

Smolen P, Baxter D, Byrne J. 2000. Mathematical modeling of gene networks. Neuron 26:567–580.

Somogyi R, Sniegoski C. 1996. Modeling the complexity of gene networks: understanding multigenic and pleiotropic regulation. Complexity 1:45–63.

Sorlie T, Perou CM, Tibshirani R, Aas T, Geisler S, Johnsen H, Hastie T, Eisen MB, van de Rijn M, Jeffrey SS, et al. 2001. Gene expression patterns of breast carcinomas distinguish tumor subclasses with clinical implications. Proc Natl Acad Sci USA 98(19):10869–10874.

Spellman PT, Rubin GM. 2002. Evidence for large domains of similarly expressed genes in the *Drosophila* genome. J Biol 1(1):5. Epub 2002 Jun 18.

Stoeckert C, Pizarro A, Manduchi E, Gibson M, Brunk B, Crabtree J, Schug J, Shen-Orr S, Overton GC. 2001. A relational schema for both array-based and SAGE gene expression experiments. Bioinformatics 17:300–308.

Stuart JM, Segal E, Koller D, Kim SK. 2003 A gene-co-expression network for global discovery of conserved genetic modules. Science 302:249–255.

Szallasi Z, Liang S. 1998. Modeling the normal and neoplastic cell cycle with realistic boolean genetic networks: their application for understanding carcinogenesis and assessing therapeutic strategies. Pacific Symp Biocomp 3:66–76.

Talamo F, D'Ambrosio C, Arena S, Del Vecchio P, Ledda L, Zehender G, Ferrara L, Scaloni A. 2003. Proteins from bovine tissues and biological fluids: defining a reference electrophoresis map for liver, kidney, muscle, plasma and red blood cells. Proteomics 3:440–460.

Tennant RW. 2002. The National Center for Toxicogenomics: using new technologies to inform mechanistic toxicology. Environ Health Perspect 110:A8–A10.

Theilhaber J, Connolly T, Roman-Roman S, Bushnell S. 2002. Finding genes in the C2C12 osteogenic pathway by k-nearest neighbor classification of expression data. Genome Res 12:165–176.

Thompson KL, Afshari C A, Amin RP, Bertram TA, Car B, Cunningham M, Kind C, Kramer JA, Lawton M, Mirsky M, et al. 2004. Identification of platform-independent gene expression markers of cisplatin nephrotoxicity. Environ Health Perspect 112:488–494.

Thomas R, Thieffry D, Kaufman M. 1995 Dynamical behaviour of biological regulatory networks - I. Biological role of feedback loops and practical use of the concept of the loop- characteristic state. Bull Math Biol 57:247–276.

Toraason M, Albertini R, Bayard S, Bigbee W, Blair A, Boffetta P, Bonassi S, Chanock S, Christiani D, Eastmond D, and others. 2004. Applying new biotechnologies to the study of occupational cancer—a workshop summary. Environ Health Perspect 112:413–416.

Travlos GS, Morris RW, Elwell MR, Duke A, Rosenblum S, Thompson MB. 1996. Frequency and relationships of clinical chemistry and liver and kidney histopathology findings in 13-week toxicity studies in rats. Toxicology 107:17–29.

Ulrich RG, Rockett JC, Gibson GG, Pettit SD. 2004. Overview of an interlaboratory collaboration on evaluating the effects of model hepatotoxicants on hepatic gene expression. Environ Health Perspect 112:423–427.

van de Vijver MJ, He YD, van't Veer LJ, Dai H, Hart AA, Voskuil DW, Schreiber GJ, Peterse JL, Roberts C, et al. 2002. A gene-expression signature as a predictor of survival in breast cancer. N Engl J Med 347:1999–2009.

van Someren EP, Wessels LFA, Reinders MJT. 2000. Linear modeling of genetic networks from experimental data Intelligent Systems for Molecular Biology (ISMB 2000), San Diego, CA, August 19–23.

van't Veer LJ, Dai H, van de Vijver MJ, He YD, Hart AA, Mao M, Peterse HL, van der Kooy K, Marton MJ, Witteveen AT, et al. 2002. Gene expression profiling predicts clinical outcome of breast cancer. Nature 415:530–536.

Vilain C, Libert F, Venet D, Costagliola S, Vassart G. 2003. Small amplified RNA-SAGE: an alternative approach to study transcriptome from limiting amount of mRNA. Nucleic Acids Res 31:e24.

Vitorino R, Lobo MJ, Ferrer-Correira AJ, Dubin JR, Tomer KB, Domingues PM, Amado FM. 2004. Identification of human whole saliva protein components using proteomics. Proteomics 4:1109–1115.

Wagenaar GTM, ter Horst SA, van Gastelen MA, Leijser LM, Mauad T, van der Velden PA, de Heer E, Hiemstra PS, Poorthuis BJ, Walther FJ. 2004. Gene expression profile and histopathology of experimental bronchoplmonary dysplasia induced by prolonged oxidative stress. Free Radical Biol Med 36:782–801.

Waring JF, Ciurlionis R, Jolly RA, Heindel M, Ulrich RG. 2001. Microarray analysis of hepatotoxins in vitro reveals a correlation between gene expression profiles and mechanisms of toxicity. Toxicol Lett 120:359–368.

Waring JF, Gum R, Morfitt D, Jolly RA, Ciurlionis R, Heindel M, Gallenberg L, Buratto B, Ulrich RG. 2002. Identifying toxic mechanisms using DNA microarrays: evidence that an experimental inhibitor of cell adhesion molecule expression signals through the aryl hydrocarbon nuclear receptor. Toxicology 181-182:537–550.

Waring JF, Ulrich RG, Flint N, Morfitt D, Kalkuhl A, Staedtler F, Lawton M, Beekman JM, Suter L. 2004. Interlaboratory evaluation of rat hepatic gene expression changes induced by methapyrilene. Environ Health Perspect 112:439–448.

Waters MD, Boorman G, et al. 2003. Systems toxicology and the chemical effects in biological systems knowledge base. Environ Health Perspect 111:811–824.

Waters MD, Olden K, Tennant RW. 2003. Toxicogenomic approach for assessing toxicant-related disease. Mutat Res 544:415–424.

Weaver DC, Workman CT, Stormo GD. 1999. Modeling regulatory networks with weight matrices. Pacific Symp Biocomp 4:112–123.

Weiss A, Delproposto J, Giroux CN. 2004. High-throughput phenotypic profiling of gene-environment interactions by quantitative growth curve analysis in Saccharomyces cerevisiae. Anal Biochem 327:23–34.

Welford SM, Gregg J, Chen E, Garrison D, Sorensen PH, Denny CT, Nelson SF. 1998 Detection of differentially expressed genes in primary tumor tissues using representational differences analysis coupled to microarray hybridization. Nucleic Acids Res 26(12):3059–3065.

Wesselman JP, Kuijs R, Hermans JJ, Janssen GM, Fazzi GE, van Essen H, Evelo CT, Struijker-Boudier HA, De Mey JG. 2004. Role of the rhoa/rho kinase system in flow-related remodeling of rat mesenteric small arteries in vivo. J Vasc Res 41:277–290.

Wolf DM, Eeckman FH. 1998. On the relationship between genomic regulatory element organization and gene regulatory dynamics. J Theor Biol 195(2):167–186.

Wuensche A. 1998. Genomic regulation modeled as a network with basins of attraction. Pacific Symp Biocomp 3:89–102.

Xirasagar S, Gustafson S, Merrick BA, Tomer KB, Stasiewicz S, Chan DD, Yost KJ 3rd, Yates JR 3rd, Sumner S, Xiao N, and others. 2004. CEBS object model for systems biology data, SysBio-OM. Bioinformatics 20(13):2004–2015. Epub 2004 Mar 25.

Yuh C-H, Bolouri H, Davidson EH. 1998. Genomic cis-regulatory logic: experimental and computational analysis of a sea urchin gene. Science 279:1896–1902.

Zhou X, Wang X, Dougherty ER. 2003a. Construction of genomic networks using mutual-information clustering and reversible-jump Markov-chain-Monte Carlo predictor design. Signal Proc 83(4):745–761.

Zhou X, Wang X, Dougherty ER. 2003b. Binarization of microarray data on the basis of mixture model. Mol Cancer Ther 2(7): 679–684.

Zmasek CM, Eddy SR. 2002. RIO: analyzing proteomes by automated phylogenomics using resampled inference of orthologs. BMC Bioinformatics 3(1):14.

Zweiger G. 1999. Knowledge discovery in gene-expression-microarray data: mining the information output of the genome. Trends Biotechnol 17:429–436.

5 The Extension of Molecular and Computational Information to Risk Assessment and Regulatory Decision Making*

*James S. Bus, Richard A. Canady, Tracy K. Collier,
J. William Owens, Syril D. Pettit, Nathaniel L.
Scholz, and Anita C. Street*

5.1 INTRODUCTION

This chapter attempts to anticipate how omics technologies could improve decisions through their application to risk assessments at present and in the future, what omics technologies are unlikely to accomplish for the foreseeable future, what systemic barriers exist for risk assessors that may inhibit or diminish the use of omics technologies (at least for the near term), and which weaknesses and shortcomings of omics technologies should be the focus of research efforts.

Most forms of toxicity, including chemically induced toxicity, begin with the alteration of molecular processes and result in effects at higher levels of biological organization (e.g., the toxicity evident in an entire organ such as liver). Classical toxicology used by risk assessors has rarely been able to obtain or to incorporate information of changes at the molecular scale. However, scientific advances now offer risk assessors the opportunity and challenge to simultaneously evaluate the expression of multiple genes and the modulation of numerous metabolic pathways. The information about the function of tens of thousands of molecularly based

---

* The findings and conclusions in this article have not been formally disseminated by the Food and Drug Administration, National Oceanic and Atmospheric Administration, or Environmental Protection Agency and should not be construed to represent any agency determination or policy.

responses comes from a family of new techniques called omics technologies. Application of these techniques to toxicology has the potential to expand our understanding of fundamental mechanisms of toxicity. This promises to improve our knowledge of the cross-species relevance of surrogate test system findings to human and environmental species, replace default assumptions in some instances, and reduce uncertainties in many risk assessments. These new tools may also be applicable for use in cumulative risk assessment where specific modes of action have been identified.

The term omics technologies broadly combines 3 areas of molecular analysis: Transcriptomics is the analysis of gene transcription molecules (messenger RNA); proteomics is the presence and possible post-transcriptional modification of proteins; and metabolomics undertakes the analysis of small molecular weight compounds that are typically metabolites of cell processes. Each technology aspires to analyze all of its respective targets in a biological sample that may be an in vitro cell culture, a tissue sample, a serum sample, or in some cases a whole organism. The challenges are then not only to achieve an understanding and integration of these technologies with each other, but also to integrate these molecular events with our understanding of classical toxicology.

A central premise for this chapter is that omics technologies have the potential to improve decisions regarding safety by reducing uncertainties in risk assessment that are, at present, difficult or intractable to address using classical methods. Two additional premises were developed during the workshop discussions: 1) Omics technologies still require significant additional development and understanding before they will achieve wide use by risk assessors, and 2) omics technologies are, at least for the near term, more likely to improve rather than replace existing testing and risk assessment methods. Whether, when, and to what degree these technologies may replace current data types or risk assessment methods will depend on the nature of the decisions being made and the degree of acceptance by stakeholders regarding the predictive power of omics technologies for human or ecological health outcomes.

Though beyond the scope of this chapter, it is also important to acknowledge the growing body of integrated and interdisciplinary research that is exploring and anticipating the implications of convergent technologies of which omics is a part. This convergence of bio-, nano-, and information technologies has the potential to expand our overall understanding of complex natural and biological systems and their function at the nanoscale. This is particularly noteworthy because an improved understanding of cellular processes could lead to significant breakthroughs in predictive toxicology, including cross-species extrapolation.

5.1.1 Scope of the Chapter

The workshop that is the basis for this publication focuses on the use of omics to address cross-species extrapolation, which is a fundamental issue and source of uncertainty for human and ecological risk assessment. Data from surrogate species are almost always applied to or extrapolated from another species by risk assessors. Human risk assessment is largely predicated on using a few laboratory animals (rats, mice, and/or rabbits for most chemicals with the addition of dogs and primates for some agrochemicals and pharmaceuticals) as surrogates for a single species: humans. Traditional ecological risk assessment, in contrast, uses a very limited number of surrogate species to extrapolate across a wide range of phylogenetically diverse organisms. Extrapolation across species within a single phylogenetic class is not uncommon, but also poses scientific challenges because of the potential for great diversity in reproductive or other life strategies, or even genetic composition, present in a given class.

Human and ecological risk assessment is confronted by numerous uncertainties, largely emanating from the need to extrapolate toxicological data across species. These extrapolations are primarily challenged by uncertainties associated with differences in the systemic and biochemical processing pathways across species. It is these mechanistic uncertainties that omics technologies offer the potential to elucidate, rather than assume, by defining species similarities and differences through global analyses at the molecular level. However, the technical unknowns and the possible paths to realizing this potential are numerous, and barriers to effective use of the technologies in risk assessment and decision-making may be substantial.

The primary objectives of this chapter are to clarify the potentially significant knowledge gains and beneficial applications that make omics technologies so attractive; discuss future expectations for omics in relation to current and anticipated practical applications of these technologies; define the technical issues that must be addressed in order to develop, gain experience in, and consider validation of omics technologies for risk assessment supporting regulation and other risk management decision arenas involving cross-species extrapolation; and address other nontechnical potential limitations to the routine application of omics for risk assessment (e.g., availability of trained risk assessors to evaluate the data, privacy concerns associated with sample procurement). This chapter is divided into a sequence of sections that describe

- the current state of practice for human and ecological risk assessment, including issues of cross-species extrapolations and the source of uncertainties;
- the relationship between statutory requirements and data development, and submission issues;
- the relationship between biological scale and assessment (addressing the overarching difficulty of linking molecular omics results with classical higher-level assessment endpoints);
- the uncertainties in applying omics technologies to human risk assessments
- the uncertainties in applying omics technologies to ecological risk assessments;
- the importance of "groundtruthing" to build evolving experience with omics technologies into the established process for linking biological response data to endpoints for human and ecological risk assessment;
- the human and equipment infrastructure limitations that could prevent risk assessors from using omics results and recommendations on how to address these system limitations; and
- the overall conclusions and recommendations.

5.2 OVERVIEW OF HUMAN AND ECOLOGICAL RISK ASSESSMENT

Risk managers are charged with the enormous and complex task of making decisions that protect human and environmental health. Risk assessments provide critical information for those decisions. The primary emphasis of risk assessment in its purest sense is to characterize quantitatively the potential adverse impacts, including associated uncertainty or likelihood, of chemical and/or other environmental stressors on human health and/or environmental species and ecosystems. Although genomics information is significantly conserved across living organisms, empirical experience has made it nonetheless abundantly evident that responses of different species to environmental stressors are often hugely varied qualitatively and quantitatively.

Our lack of understanding of this variation necessitates the use of default assumptions to fill in gaps when extrapolating across species. These default assumptions are applied in what is assumed to be a protective direction for the stressors considered so that humans or critical species are assumed to be more sensitive than the tested laboratory animals or surrogate species. While historical methods of biology, toxicology, and environmental sciences are indeed capable of providing scientific information useful for risk assessment, the advent of omics technologies offers promising future opportunities to evaluate directly critical risk assessment assumptions underpinning cross-species extrapolations. Reduction of these default assumptions, their application factors, and other uncertainties as a result of an improved understanding of dose–response relationships will improve science-based risk assessments and the decisions that rely on them.

Science-based inputs that reduce the inherent uncertainties of risk assessments would have substantial societal benefits, including direct effects on health and well-being. Poorly characterized uncertainties resulting in overly conservative risk evaluations can deny society important benefits from loss of valued products or services and introduce large costs that carry minimal health or environmental gains. These lost opportunities and costs potentially have secondary effects on productivity, investment, and quality of life. However, perhaps more importantly, the application of limited public resources toward stressors that are not causing harm means that those resources are not being optimally applied to problems caused by agents known to cause harm. Similarly, use of assumptions that lead to underestimation of true risks may have devastating cost and health consequences to society. Thus, technologies and science that confidently and validly reduce risk assessment uncertainties offer huge benefits to society.

One probable way to address poorly characterized uncertainties described here is to develop a coherent, comprehensive statement of risk through an integrated risk assessment. Such assessments would likely result in a more effective, efficient, holistic approach while providing a more complete picture of overall ecosystem and human health and a stronger basis to support sound decision-making. Assessments that do not integrate health and ecological risks, as currently practiced, are likely to miss important modes of action that involve interactions between effects on the environment and effects on humans. There is a growing recognition of the interdependence of these 2 disciplines and the need to better protect humans as well as the environment; this suggests the need to develop an approach to risk assessment that addresses situations of multichemical, multimedia, multiroute, and multispecies exposures in a more realistic manner (Suter et al. 2001).

As for the advent of omics technologies, any benefits or drawbacks as a result of their application to integrated risk assessments will be similar in nature and scope to the currently decoupled approaches. However, any drawbacks may be over-shadowed by the benefits to unifying risk approaches. Clearly, omics will likely lead to tremendous breakthroughs in our understanding of mechanistic action in toxicological events and the identification of specific biomarkers of exposure, injury, or susceptibility. However, the routine use of omics in any context will be largely dependent on the ability to link toxicogenomics to disease outcomes and adverse effects.

5.2.1 HUMAN HEALTH RISK ASSESSMENT

The fundamental approach to human and environmental risk assessment has been defined as the coupling of hazard (dose–response) characterization information obtained from surrogate species representing target populations with estimated exposure data to environmental stressors (NRC 1983). Potential applications of omics technologies to hazard and exposure will be identified in the following sections.

For most environmental, occupational, or food-related human health assessments, existing risk evaluation practice requires hazard characterization testing that seeks to identify the range and dose response of potential adverse responses associated with specific chemical treatments in animal surrogates. The primary objective of these studies is to establish an overall highest no observed adverse effect level (NOAEL) dose (associated with the lowest observed adverse effect level or LOAEL) for the most sensitive endpoint (e.g., neurotoxicity, liver damage) in the most sensitive species tested (e.g., rat, mouse, rabbit). The NOAEL value is assumed to represent a dose at which adverse outcomes are not likely to be present for the specific endpoints evaluated in the species studied. For all noncancer and some cancer (nongenotoxic) responses, human exposure doses at which adverse outcomes are not likely are derived by sequentially dividing the appropriate overall animal NOAEL value by each of 2 primary 10-fold default uncertainty factors.

The first 10-fold value is intended to represent interspecies variation in adverse response (extrapolation from animal to human; test animals are generally assumed less sensitive than humans). The second 10-fold value represents potential variability of human responses accounted for by age, race, gender, and other biological differences (Lehman and Fitzhugh 1954). In standard regulatory practice, it is often prohibitively difficult to make a successful case that the test animal is equal to or more sensitive than humans in order to avoid use of the default 10-fold reduction for interspecies variation. In recent years, Federal legislation such as the 1996 Food Quality Protection Act (FQPA) has mandated consideration of up to an additional 10-fold uncertainty factor to represent a range in sensitivity across life stages. Thus, reasonable protection of humans from the risks of environmental contaminants is generally assumed when the human exposures are 100 to 1000 times below the lowest overall animal NOAEL values.

The result of this approach is a regulatory definition of a dose or exposure at which adverse health outcomes are not likely in humans. Depending on the regulatory context, these are termed a reference dose (RfD), reference concentration (RfC), or acceptable daily intake (ADI). The approach is intended to bias risk management conservatively across chemicals so that it is very unlikely that exposure to an RfD or ADI dose would cause health effects in humans. When the actual or estimated human dose is above the RfD or ADI, risk is judged to be unacceptable and risk management ensues. The degree to which the estimated human dose is below the RfD is important. Although the long-term risk is technically acceptable when the dosage is near the RfD, risk management actions might be taken or chemicals with higher ratios of exposure dose to RfD in screening risk assessments might be given priority in the development of data for refined risk assessment.

Furthermore, analysis of mixtures of chemicals will often involve decisions of whether to add exposures below the RfD, and in some cases even refined risk assessments are forced to use default assumptions about additivity, which add further conservatism. Because of this protective bias built into their derivation and use, exceeding the RfD in the short term or even the long term does not always mean, however, that an effect will occur. Therefore this safety assessment approach does not express risk of disease incidence in a quantitative way.

There are alternative approaches such as margin of exposure (MOE) analysis that use ratios of the exposures of interest to the LOAEL or other measures of effective dose in order to develop expressions of relative proximity of environmental or food exposure levels to those that may cause adverse effects. Similar to safety assessment approaches, MOEs are evaluated during risk management using consideration of inter- and intraspecies variability in susceptibility. Decisions using MOE approaches are less constrained to a specific number for safe exposure across exposure scenarios, at least in theory.

Each of the uncertainty factors described here is subject to refinement when data are available that clearly reduce the uncertainty. For example, better understanding of animal–human cross-species extrapolation or better understanding of human response variability can lead to a reduction in the default 10-fold values. For example, use of pharmacokinetic modeling and dose metrics that remove some of the interspecies extrapolation concerns can reduce the interspecies extrapolation factor to 3-fold, rather than 10-fold.

A different approach is taken for environmental agents that are suspected to cause cancer in animals by mechanisms found to damage genetic material directly (i.e., genotoxic agents). Such substances are assumed to have no threshold for risk, and thus cancer risks are calculated from models using linear assumptions of animal cancer dose–response data. Generally, population-based risk levels in the range of 1 cancer case per 10,000 to 1 cancer case in 10 million exposed individuals are the de minimis risk targets for US regulatory decisions. In these cases, precautionary assumptions about interspecies extrapolation also occur, albeit not through the use of uncertainty factors. For example, in the absence of mechanistic information that would argue to the contrary, any tumor stage or type observed in animals is considered a relevant risk to humans. In fact, these tumorigenic responses are known to vary by species and some test species are known to be more sensitive to some types of cancer than humans are or even uniquely sensitive to the cancer.

In general, the central assumption of human risk assessment is that the toxicologic responses observed in limited numbers of animal test species are indeed predictive of potential adverse human health effects. Decades of empirical experience comparing animal bioassay responses to those seen in humans encountering relatively high doses of environmental substances (therapeutic agents, high-level occupational exposures, accidents) demonstrate some predictive ability at high doses. However, when human exposures are significantly below the doses eliciting responses in animal test systems (as is often the case with residual environmental chemical exposures), then mode and/or mechanism of action, dose-dependent transitions in mechanisms of action, cross-species extrapolation, and other critical biologic factors become far more useful determinants in estimating actual human risk.

Significant opportunities exist to provide better precision and confidence in estimating health risks under conditions of low-level chemical exposures commonly encountered under real-world environmental conditions. This essential requirement is most often vexed by an inability to correlate the molecular events ultimately leading to cancer with lower doses and by cross-species extrapolation issues. The lack of a clear understanding of the biological similarities and differences between animals and humans stands as the central barrier to moving beyond simple use of scientifically undefined 10-fold uncertainty factors or assumptions of low-dose linearity and understanding and accurately assessing human risk. If implemented successfully, the use of omics approaches has the potential to provide clarity into mechanistic differences in toxicological responses across species.

5.2.2 Ecological Assessment

With respect to environmental contamination, the typical aim of an ecological risk assessment is to estimate the potential impacts of pollutants on at-risk organisms, populations, species, communities, or ecosystems (Figure 5.1). Similar to assessments for human health, risk is generally characterized in terms of stressor–response, or the likelihood that an ecological exposure will result in an adverse toxicological outcome. Thus, the risk = hazard × exposure equation is also the simple premise underlying many current applications of ecological risk assessment (USEPA 1998).

Although human health and ecological risk assessments share the common goal of defining and refining the stressor–response relationship, they often differ in scope, spatial and temporal scale, quantitative precision, and complexity. For humans, the risk question is usually focused narrowly on the health of individuals. Also, relative to ecological receptors, human health risk assessments are typically rich in data. For example, mammalian toxicity studies tend to be more numerous than studies on other nonmammalian taxa, and they also tend to be more mechanistic. As a consequence, key uncertainties in risk assessments for humans are usually well defined and narrow in scope (i.e., the extent to which results from one mammalian species can be extrapolated to another).

With the exception of threatened or endangered species, ecological risk assessments rarely focus on the health of individual organisms alone. Hazard is more

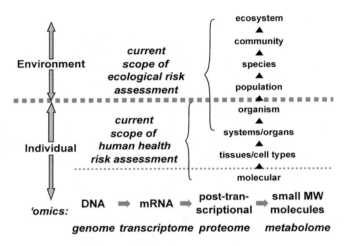

FIGURE 5.1 Biological scale is an important consideration when evaluating omics information for the purposes of human health and ecological risk assessment. Emerging omics technologies are designed to monitor chemically induced responses at the molecular level. A recurring challenge in risk assessment is to relate these molecular responses to toxicological outcomes at higher scales of biological complexity.

commonly defined as adverse impacts at higher scales of biological organization, including populations, communities, or ecosystems. Overall, the problem of extrapolation then becomes much more complex for ecological risk assessments. This includes 1) cross-species extrapolations, 2) extrapolations from suborganismal toxicity data to populations and communities, 3) extrapolations that vary in space and time (i.e., exposures and ecological responses that vary across landscapes), and 4) extrapolations from simplified laboratory environments to complex natural systems. For these and other reasons, ecological risk assessments tend to be more complex, less quantitative, and less mechanism-oriented than risk assessments for human health.

At present, toxicity data from a few model test organisms are often used to estimate risk to hundreds or even thousands of species. For example, the USEPA uses a tiered approach to evaluate pesticide registrations under the US Federal Insecticide, Fungicide, and Rodenticide Act (FIFRA). The minimum ecological effects data required to support the registration of an outdoor-use pesticide include 6 basic studies (Touart and Maciorowski 1997):

1) an avian single-dose oral toxicity test,
2) an avian dietary toxicity test with an upland gamebird,
3) an avian dietary test with a waterfowl,
4) a freshwater fish acute toxicity study with a warm-water species,
5) a freshwater fish acute toxicity test with a cold-water species, and
6) an aquatic invertebrate acute toxicity test with an immature life stage.

Data for many large taxonomic groups are not required and must therefore be extrapolated from the preceding model species (i.e., using acute data from rainbow trout to estimate toxicity to amphibians). Recently, omics approaches have advanced in a wide array of species (see Chapter 3). This raises the possibility of comparative studies to evaluate the relative toxicological responses of different species to pesticides and other environmental contaminants. While species-specific differences are inevitable, a more detailed understanding of toxicity across species should help reduce a key source of uncertainty in ecological risk assessments.

Acute lethality is one of the more common endpoints used in ecological risk assessments and is typically expressed as the 96-hour LD50 or LC50 (median lethal dose or concentration, respectively). In the preceding example of a tier-1 screen under FIFRA, ecological risk is assessed by comparing acute toxicity data for a pesticide to that chemical's estimated environmental concentration (EEC). In the context of FIFRA, additional review may be warranted if the EEC exceeds some predetermined fraction (i.e., 1/10) of the LC50 for an aquatic species of concern. Risk assessments may also estimate exposure using probabilistic methods and ecological response using species-sensitivity distributions, which are typically the LC50s of several species that may (or may not) comprise a community ensemble.

Irrespective of the approach, the underlying assumption that ecological responses can be predicted from laboratory-derived LC50s is a subject of continuing debate. At issue is whether safety factors of 10, 100, or even 1000 (in the case of the European Union) are protective when applied to acute data. There are several sources of uncertainty. First, large-scale and long-term ecotoxicological investigations have recently highlighted the importance of sublethal effects in terms of the impacts of pollution on natural populations (Peterson et al. 2003). Second, the median lethal concentration may miss important temporal effects, including delayed mortality (Zhao and Newman 2004). Third, individually based LC50 measures may fail to capture important, species-specific demographic life history characteristics, such as life span, time to reproduction, and number of offspring produced. This can place limits the usefulness of LC50 data derived from model (i.e., laboratory) test organisms, particularly in terms of predicting the susceptibility of native species that have different life histories (Stark et al. 2004). Fourth, single-species distributions (using LC50s or other endpoints) often lack ecological realism because the distributions are based on the available toxicological data and not on collections of species that represent actual communities (Forbes and Calow 2002). Finally, ecological risk assessments based on LC50 data for a few species may fail to address cascading indirect effects that influence important aspects of ecosystem structure and function (Preston 2002).

Thus, at present, at least 2 major challenges confront the field of ecological risk assessment. The first is to capture a greater degree of ecological realism, particularly during the critical phase of problem formulation (USEPA 1998). This will necessitate an increasingly multidisciplinary approach involving toxicologists, population and community ecologists, and conservation biologists (Hansen and Johnson 1999). A second challenge is to critically assess acute mortality as a tier-1 ecotoxicological endpoint. Specifically, mechanisms of toxicity need to be elucidated at scales below the level of the individual organism (i.e., sublethal) and, at the same time, linked and correlated to biological process that ultimately determine the survival, reproductive success, and distribution of individuals within populations and communities.

From there, it will be important to determine the extent to which known pathways of toxicity and modes of action are conserved in laboratory versus wild species. Accordingly, future mechanism-oriented research in ecotoxicology should benefit considerably from the advancement of omics technologies.

5.2.3 DIFFERENCES IN STATUTORY REQUIREMENTS

A long-held adage in the field of toxicology is that "the dose makes the poison." This reflects the central importance of the dose–response relationship in evaluating a chemical's potential hazard in terms of a biological assessment endpoint or suite of endpoints. Following the paradigm, this hazard is then combined with the exposure potential to estimate risk. While the dose may make the poison, it is important to note that often the statute makes the risk by defining the hazard, exposure, or level of risk for circumstances addressed by the statute. FQPA is one example in allowing an extra uncertainty factor to be applied in the case of children. However, compared to relatively uniform approaches to human health risk, there are diverse answers to the question of what constitutes ecological risk because the ecological risk assessments are typically tailored to, and constrained by, the unique statutory or regulatory context for asking the natural resource management question in the first place. This means that different statutes afford different levels of protection for ecological receptors that range from individual animals to ecosystems.

5.2.3.1 Human Health Risk Assessment

The level of proof needed to establish that harm may occur for a decision supported by human health risk assessment varies greatly across national regulatory jurisdictions and also across statutes and even across separate phases or applications within a statute with a given jurisdiction. For example, screening level assessments for site or contaminant prioritization within USEPA Superfund programs can be based on much less data than for the eventual record of decisions for individual sites and much less than for registration of pesticides. Similarly, for the US Food and Drug Administration (USFDA), drug or food additive approvals require extensive datasets and analysis rigor compared to that currently required for dietary supplements or cosmetics.

5.2.3.2 Ecological Risk Assessment

The degree of protection afforded to plants, fish, wildlife, and other biological organisms varies considerably according to the regulatory context for natural resource decision-making. Using the United States as an example, several different Federal statutes address the potential ecotoxicological impacts of chemical contaminants on native, nontarget species. In addition to the earlier example of pesticide registration under FIFRA, other significant statutes include the Endangered Species Act (ESA), the National Environmental Policy Act (NEPA), the Comprehensive Environmental Response, Compensation, and Liability Act (CERCLA, a.k.a. Superfund), the Clean Water Act (CWA), the Migratory Bird Treaty Act (MBTA), the Marine Mammal Protection Act (MMPA), essential fish habitat provisions of the

Magnuson–Stevens Fishery Conservation and Management Act (EFH), the Bald and Golden Eagle Protection Act (BGEPA), and the Oil Pollution Act (OPA).

The narrative standards for these different regulatory acts range from "no unreasonable impacts to nontarget species" (FIFRA) to "no toxics in toxic amounts" (CWA) to "no harm to individual animals" (ESA). Also, depending on the environmental decision-making context, the question of risk may range from individual animals (MMPA) to populations (FIFRA), species (ESA), communities (CWA), or ecosystems (OPA). For these reasons, the scope and complexity of ecological risk assessments can vary considerably. It can also be difficult to harmonize risk assessments in situations where distinct statutes overlap in authority (i.e., FIFRA, CWA, and ESA all have application to pesticide registration).

This chapter is largely focused on the technical applications of omics methods to reduce uncertainties in cross-species extrapolations and, by extension, ecological risk assessment. Because omics technologies are almost entirely empirical, the uncertainties addressed by omics will be largely epistemic (uncertainty in determinate facts; Regan et al. 2002). These include familiar sources of uncertainty such as subjective judgment, measurement error, natural variation, and model uncertainty. It is important to note that epistemic uncertainty is distinct from linguistic (or language) uncertainty, which often pervades ecological risk assessments. Vagueness, context dependence, and underspecificity are common sources of linguistic uncertainty (Regan et al. 2002). For example, in the preceding paragraph, what specific toxicological responses might constitute "unreasonable impacts," "toxic amounts," or "harm"? Advances in omics will do little to reduce linguistic uncertainties in ecological risk assessment because these typically have their origins in the language of different statutes.

Depending on the regulatory context, a risk assessment might address the potential for injury to an individual marine mammal or it might evaluate the likelihood that the trophic structure of a lake ecosystem will collapse as a result of continuous inputs of chemical contaminants. However, no matter the biological scope of the risk assessment, the initial and most critical phase of an ecological risk assessment is problem formulation (USEPA 1998). During problem formulation, the regulatory context is identified, risk hypotheses are defined, and a conceptual model is developed.

A key step is the selection of assessment endpoints, which are broadly defined as "explicit expressions of the actual environmental value that is to be protected, operationally defined by an ecological entity and its attributes" (USEPA 1998). For traditional ecological risk assessment, a near-term challenge will be to link subtle molecular changes (as revealed by omics technologies) to conventional whole-organism endpoints such as growth, mortality, and reproduction. However, other assessment endpoints may also be useful in terms of linking omics data to risk assessment. These endpoints can be intraspecific (e.g., endocrine disruption, migratory behavior) as well as interspecific (e.g., disease susceptibility, predation vulnerability). Moreover, there is considerable potential for omics approaches to improve the diagnostic capabilities of field investigations. This includes improved and more specific biomarkers of exposure as well as molecular bioindicators of adverse toxicological response.

5.3 POTENTIAL OF OMICS TO IMPROVE RISK ASSESSMENT

5.3.1 REDUCING UNCERTAINTY IN HUMAN HEALTH RISK ASSESSMENT

5.3.1.1 Omics Approaches to Addressing Cross-Species Uncertainties

Omics technologies offer a number of potentially attractive approaches to reduce the uncertainties associated with cross-species extrapolation using high doses of a substance in animal toxicity tests to estimates of human risk, where humans are exposed to much lower environmental levels of the same substance. However, the use of omics technologies in risk assessment practice must be implemented with caution. In the absence of strategic development of cross-species validation datasets, the potential for omics data to create additional undefined over- or underestimates of hazard and risk projections is high. Thus, omics data offer particular promise, as well as warranting caveats, to several important research inputs of the risk assessment process.

5.3.1.2 Screening

Screening assays are intended to rapidly identify potential intrinsic hazard characteristics of chemicals (e.g., genotoxicity, cytotoxicity, endocrine activity) in order to provide a reasonable rationale for triggering more resource-intensive apical whole-animal bioassays. Screening assays are most useful for setting research testing priorities from among larger groups of chemicals. These assays, which can be in vitro or in vivo, provide only rough estimates of health hazards (potential to cause injury). They generally cannot be used to infer any actual health risks.

The benefits of screening assays in reducing cross-species uncertainties in risk assessment are largely indirect. For any given specific endpoint (e.g., hepatotoxicity), omics-based assays may allow more rapid identification of the appropriate animal model to use in subsequent apical animal tests, assuring that the selected animal model is most likely to represent human responsiveness. In addition, gene expression patterns, protein modification, or metabolites may provide early insights into potential mechanisms of toxicity. Such information can signal potential needs for special study designs or specific attention to specific target organs. Finally, of course, identification of mechanisms provides insight on priorities for moving individual compounds from among a group or class of compounds to more definitive apical tests, from which the risk assessor is more likely to receive important, detailed dose–response and target organ data on endpoints of critical interest.

However, it is important to realize that omics screening data likely will not be able to provide insights into many important cross-species issues that have an impact on extrapolation of animal data to humans. In many cases, genomics screens will need to be supplemented by metabolomics or other data in order to provide information on critical interspecies pharmacokinetic differences important to risk extrapolation. For example, the widely used organic acid herbicide 2,4-dichlorophenoxy acetic acid (2,4-D) is significantly more toxic to dogs than to rodents. This toxicity difference is attributable to the fact that dogs are unable to clear 2,4-D rapidly due to a lack of a renal organic anion active transport system. Rodents and humans possess the renal anion transport clearance system and rapidly clear 2,4-D from the body after oral absorption (Timchalk 2004).

The cross-species difference in renal transport accounts for the observation that plasma levels of 2,4-D are significantly higher in dogs compared to rats and humans administered equivalent oral doses. This species sensitivity of dogs relative to rats and humans applies not only to 2,4-D, but also to many other low molecular weight organic anion toxicants. In the absence of any other toxicity and metabolic information, a priori screens of transcribed genes would fail to identify such critical pharmacokinetic toxicity drivers. There are many examples of other cross-species pharmacokinetic differences that also would similarly go unidentified in a strictly genomics screening protocol.

Proteomics and metabolomics (but not transcriptomics) screening of serum, urine, and tissues may provide information on a mode of action as well as potential target organs and critical suborgan targets that might result from intact animal dosing. In fact, extended whole-animal dosing with multiple-organ transcriptomics screening in multiple test species will almost certainly prove logistically and cost prohibitive. However, if preliminary whole-animal toxicity screening data or specific proteomics and metabolomics data reveal potential target organs in a single model species, then subsequent transcriptomics screening tests in other species might be identified and selected for the purposes of cross-species extrapolation, as well as for the development of structural activity relationships for related substances.

For many reasons similar to those described here, it appears unlikely that introduction of omics methods into toxicology will prove sufficient to replace traditional intact animal toxicity testing fully. However, opportunities for its use to complement and improve traditional toxicity testing in specific circumstances are likely to emerge in the near term. Reductions in animal use may occur as such information is able to identify which animal species represent the most appropriate human surrogate, potentially eliminating need for traditional multiple-species testing.

5.3.1.3 Impact of Genomics Technology on Reducing Uncertainty in Chemical-Specific Risk Assessments

Conventional multispecies toxicity testing often reveals cross-species differences in toxicity expression in terms of dose response as well as target organs. The critical question for risk assessment to address is which species is most likely the relevant surrogate for humans; in the absence of ancillary data, the most sensitive species is always assumed to represent human responsiveness. Resolution of interspecies response issues can have enormous impact on the choice and design of toxicological studies and in risk assessment outcomes, and consequently on societal costs and benefits.

This common scenario is exemplified with the dramatic multispecies differences seen in responsiveness to treatment with dioxin-like substances. Some rat strains are extremely responsive to dioxin exposure, while other rat strains are relatively resistant, as are other animal species such as hamsters. Direct acute and chronic human experience with dioxin exposures also suggests similar resistance to dioxin and possibly differential target organ responsiveness as well. However, in the absence of clear ancillary mechanism information defining these differences, current dioxin risk assessment is defaulted to data from a relatively sensitive strain of rat. Omics methods offer great promise for breaking the mechanistic code that holds the information of which animal responses are most applicable for estimating human risks to dioxin exposure. New methods are also promising in terms of assessing the relevance of animal data to human risk in other well-studied circumstances.

5.3.1.4 Use of Genomics Methods for Refining Operational Principles of Risk Assessment

Regulatory agencies commonly issue guidance documents that proscribe important elements in the design of toxicology tests, as well as develop policy positions on how the results of toxicology information are translated to human risk evaluation. These documents reflect agency policy on how results from animal testing are extrapolated to potential human risk. Examples of such policy guidance are illustrated next.

Dose selection. Responses observed in animal toxicity studies are assumed to represent potential health hazards in the broad human population. However, animal studies practically can use only a small number of animals — a potential key weakness in extrapolating animal findings to humans. To compensate for this experimental limitation in identifying potential health hazards, regulatory agencies have required that the highest level dose used in toxicology studies achieve what is termed the "maximum tolerated dose" (MTD). The MTD must be a dose sufficiently high to produce some evidence of toxicity in the animals but not likely to adversely affect survival in long-term bioassays. In most cases, achievement of an MTD is ascertained by relatively simple measures such as a 10% depression in body weight gain or evidence of mild histopathologic lesions in target tissues.

In recent years, research in toxicology has frequently demonstrated that toxicity findings observed at the MTD may not be relevant for extrapolation to human risk associated with low environmental exposures. Examples of the many potential mechanistic reasons for why such high-dose findings may not be relevant to low-dose human risks have been delineated (Dellarco 2004) and often include high-dose nonlinearities in pharmacokinetic activation, detoxification, and clearance processes. The significance of such observations for risk assessment has been emphasized by the Society of Toxicology (SOT) Risk Assessment Task Force's conclusion that "toxicological research characterized by the use of excessively high dose relative to expected human exposures and routes of administration is unlikely to be relevant to assessment of potential hazards at environmental levels of exposures" (Conolly et al. 1999).

Importantly, SOT also stressed that the clear limitations associated with use of the MTD can be largely compensated for by a better understanding of toxicity mechanisms. Mechanism data offer not only the opportunity to identify and interpret the human health relevance of potential high-dose nonlinear toxicity responses retroactively, but also promise a priori approaches to improved design of toxicity test protocols that avoid complications of nonlinear responses. Compared to existing pharmacokinetic or biochemical approaches, omics approaches may provide more rapid and improved mechanism information to guide the design and interpretation of toxicity standardized regulatory testing protocols.

Chemical mixtures. Animal hazard identification protocols have traditionally focused on single chemical evaluations, leading to single chemical risk assessments. However, in the real world, exposure to mixtures of environmental chemicals is far more representative of actual chemical exposures. This presents the challenge of how to add the potential combined toxicity hazard of low-dose exposures to complex mixtures (Wilkinson et al. 2000). Two primary approaches focusing on dose addition or response addition have been described.

Targeted omics studies may prove particularly useful in refining the current default assumptions that underpin risk assessment of mixtures. For example, the human health risk estimates used for risk management of environmental mixtures of polychlorinated dioxins (PCDDs) and structurally related compounds such as polychlorinated biphenyls (PCBs) and polychlorinated furans (PCDFs) currently include the assumption that the toxicity of all these substances is mediated through the central mechanism of activation of the nuclear hormone aryl hydrocarbon (Ah) receptor. Thus, congeners in a mixture are assigned toxic equivalency factors (TEFs) related back to baseline toxicity of 2,3,7,8-tetrachloro-dibenzo-*p*-dioxin (TCDD), a model congener.

The fundamental assumption of the TEF approach is that all congeners in the mixture are toxicologically and risk equivalent due to their common mode of action mediated through activation of the Ah receptor. Importantly, genomics approaches offer realistic opportunities to examine the validity and implications of this assumption experimentally, keeping in mind that various compounds may have toxicities other than those mediated by the Ah receptor. If in fact dioxin-like congeners are truly equivalent, as the TEF approach assumes, they should produce a subset of Ah receptor-mediated genomics expression patterns that are mostly similar. This critical assumption is already being examined.

Low-dose linearity of genotoxic carcinogens. For chemicals identified as being genotoxic, positive tumorigenic responses in animal cancer bioassays are assumed to have no threshold for response — that is, any dose, no matter how small, presents a potential cancer risk. This central assumption of no threshold for genotoxic carcinogens can carry significant societal cost and health implications when the costs and resources for achieving environmental exposure targets based on no-threshold, linear cancer assumptions are large. However, the biologic plausibility of this fundamental assumption has been challenged, primarily based on animal and human experiments with acknowledged limitations in experimental power. By potentially revealing fundamental responses of cells and organisms to low-level chemical exposures, genomics approaches may provide valuable opportunities to explore the mechanistic rationality of this no-threshold default assumption for specific genotoxins or for identifiable classes of genotoxins. Such gene expression and other changes must be evaluated in the context of overall statistical significance and with respect to background or normal fluctuations in expression.

Extending dose–response into the unobserved range. Low-dose extrapolation from classical adverse endpoints in regulatory risk assessment is typically based on a number of assumptions that may be amenable to exploration by omics techniques. For example, it is possible that omics data will be used as a point of departure (POD) for regulatory dose–response assessment, such as in developing an RfD or ADI, or in cancer assessment. One challenge of this use is that the transference of the current mindset based on LOAELs and NOAELs for measured adverse effects to the new way of extending the dose–response range will be difficult to accomplish. This issue is likely to be similar to the current argument about whether a benchmark dose low 10 (BMDL10) can be used instead of an LOAEL or NOAEL in risk assessment.

Weight of evidence. Weight-of-evidence considerations in risk assessment involve the evaluation of a full spectrum of relevant and available data. In other words, the weight of evidence requires the evaluation of all available data from all relevant disciplines: toxicology, biology, chemistry, epidemiology, statistics, etc. It should be noted that no single study drives the overall weight of evidence determination. Typically, best professional judgment regarding data adequacy, consistency, and quality is used to render a decision. Theoretically, omics data could prove to be useful in shedding light on mode of action and mechanisms for chemical toxicity; however, given the limited endpoints for the purposes of cross-species extrapolation, use of the data could be limited.

5.3.2 Reducing Uncertainty in Ecological Risk Assessment

As discussed previously, ecological risk assessments typically contend with 2 major categories of uncertainty. The first is linguistic uncertainty, which usually arises from vagueness or underspecificity in statutory language (i.e., FIFRA's "no unreasonable adverse affects to nontarget wildlife"). The second is epistemic uncertainty, or uncertainty about determinate facts (Regan et al. 2002). New omics technologies hold considerable promise for reducing several important sources of epistemic uncertainty in ecological risk assessment. In many cases, the key data gaps are similar to those for human health assessments. Omics-driven advances are likely to include, for example,

- improved mechanistic information for specific contaminants or classes of contaminants,
- the discovery of novel pathways of toxicity,
- more sensitive and mechanism-oriented screens for sublethal toxicological response (in vitro and in vivo),
- new data linking molecular responses to the fitness of individual animals (e.g., endocrine disruption and reproductive success),
- new information on the suitability of laboratory models for extrapolation to a diversity of wild species,
- new approaches to complex ecological problems such as the toxicity of chemical mixtures, and
- improved biomarkers of exposure and response (toxicological effect) for wild species in ecological systems.

The contributions of omics technologies to any of these areas will ultimately reduce uncertainty in ecological risk assessment.

Ecological risk assessments for chemical contaminants are generally poor in data. Therefore, estimating the potential sublethal impacts of pollution on the health of native species in natural systems is often problematic. In the near term, omics technologies are likely to have the greatest impact in terms of reducing uncertainty at biological scales leading up to the level of the individual organism. Much of the initial focus will be on laboratory models that have traditionally been used for toxicity testing, particularly as omics technologies are developed for these species (i.e., rainbow trout). Parallel advances can also be expected as species that are current models for omics research are increasingly incorporated into toxicological investigations (e.g., zebrafish) (Snape et al. 2004).

Omics approaches in these model organisms will be useful for addressing specific sources of uncertainty. For example, the median lethal dose or concentration (LD50/LC50) has been a predominant endpoint in the field of ecotoxicology for decades. As noted earlier, several key uncertainties are associated with the use of acute lethality data in ecological risk assessments. Also, because the LC50 and the mortality-based lowest observable effect concentration (LOEC) or no observable effect concentration (NOEC) convey little information about mechanisms of toxicity, they are limited in terms of their use for exploring the validity of high- to low-dose extrapolations and other operational principles of risk assessment. Although wildlife and fish kills are occasionally documented, the vast majority of contaminant exposures under natural conditions are sublethal. There is therefore a need for mechanistic, low-dose (i.e., sublethal) toxicological data from laboratory models where omics approaches are currently practicable. In the future, these laboratory models should provide the basis for improved toxicity screening as well as studies to elucidate fundamental pathways of toxicity that integrate responses from the molecular scale to systems physiology. Omics approaches under controlled laboratory conditions also hold considerable promise for addressing the combinatorial toxicity of chemical mixtures, the impacts of multiple stressors (chemical and nonchemical), and the extent to which toxicological mechanisms are conserved across species.

In addition to laboratory studies using model test organisms, there is an emerging role for omics technologies in field investigations. For example, a genomics approach has recently been used to evaluate the transcriptional profiles of stress-related genes from marine flatfish collected from polluted and unpolluted sites in the United Kingdom (Williams et al. 2003). It is very likely that omics will rapidly expand the toolbox of available biomarkers. These new tools have the potential to extend and improve traditional biomarker approaches, including single-gene analyses, histopathology, enzyme biochemistry, and immunohistochemistry. As with the example of marine flatfish, the initial application of omics technologies to natural systems will largely be to develop more sensitive and more specific biomarkers of exposure in non-model organisms. Subsequently, as pathways of toxicity are increasingly understood, it may be possible to develop and validate omics-related biomarkers of effect in at-risk species.

Finally, it should be reiterated that a long-standing challenge in ecotoxicology is to connect mechanisms of toxicity at scales below the level of the individual animal to toxicological outcomes at the scale of populations or higher (Figure 5.1). As a general example, the impacts of pollutants on the types of species-species interactions (i.e., competition and predation) that typically structure ecological communities are still very poorly understood. The extent to which omics technologies will reduce uncertainties at these higher scales of biological complexity is presently unclear. Moreover, many of the issues identified earlier for human health, such as interspecies variation in uptake and elimination kinetics, apply to ecological receptors as well. These will continue to pose important challenges for cross-species extrapolations in ecological risk assessment.

5.4 THE PATH FORWARD?

5.4.1 EXTRAPOLATIONS AND INFERENCES FROM OMICS DATA

One of the greatest challenges confronting risk assessors is the problem of extrapolating empirical observations at one scale of biological organization (i.e., molecular changes in mRNA levels) to higher scales (i.e., tissue and individual levels). This is particularly true for omics data. For example, an increase (or decrease) in the transcription of a particular gene does not necessarily forecast a change in translation. Similarly, changes at the level of the proteome may not necessarily lead to changes in the cellular or systems physiology of the organism. This is further complicated by the fact that toxicants may act via mechanisms and time courses that are, at least initially, independent of transcriptional processes (Miracle et al. 2003). Risk hypotheses are rarely (if ever) framed in terms of molecular responses alone. Rather, the emphasis is usually on health, which may be further defined as one or more assessment endpoints that integrate a toxicological response across molecular, cellular, tissue, and physiological scales within an individual organism (Pennie et al. 2000). For these reasons, omics approaches will be most effective in terms of reducing uncertainty in risk assessment if they 1) complement and extend traditional experimental tools for mechanism-oriented toxicology, 2) evaluate toxicological responses across multiple biological scales, and 3) are designed to test hypotheses explicitly linked to health at the level of the organism.

5.4.2 GROUNDTRUTHING AND VALIDATION

5.4.2.1 Conceptualizing Omics in the Regulatory Risk Framework

Central questions confronting risk assessors for the use of omics data concern what the data tell us and where and how omics data can be used with confidence to reduce existing uncertainties. Current risk assessments typically use assessment endpoints at organizational levels higher than that represented by omics data. Therefore, these endpoints are clearly adverse effects such as reproductive failure or arguably more clearly grounded or linked to adverse outcomes. Furthermore, the analysis and evaluation techniques for these higher organizational levels (e.g., pathology and its interpretation) are well established within regulatory risk assessment. For purposes of exploring groundtruthing and validation of omics data in a regulatory context, it is useful to divide the discussion into 3 simplistic stages:

1) Screening involving the least correlative prediction sets but extending through to mechanistic indicators
2) Weight of evidence for regulatory decisions — for example, where the omics data are submitted to supplement a new drug application or to test data from required test guideline or to define a chemical as a thresholded carcinogen or to conclude that toxicity in animal models is not relevant for humans
3) Formal test guideline submissions such as a revision of current in vitro or in vivo genotoxicity assays to discriminate between direct and indirect genotoxins

Because use of omics data from individuals in clinical and medical settings to identify likely efficacy for a drug does not rely on cross-species extrapolations, these and similar applications are not addressed here.

For the first 2 stages (use of omics to support a regulatory decision or claim), the omics data and the systems used to generate and analyze the data must be "groundtruthed." That is, parties need to demonstrate sufficient experience and understanding of the omics methods and the interpretation of their data in order to support any conclusions drawn from these data in a given regulatory context. Inherently, the degree of experience and understanding needed will rise and fall on a case-by-case basis. For the third stage (where omics testing is a required component of a regulatory testing guideline or classification scheme), the omics systems would necessarily go through more formal and proscribed demonstration of their reliability (i.e., reproducibility from lab to lab) and relevance (i.e., established link to adverse health outcome) before regulatory acceptance of their use.

The process of developing a rational groundtruthing framework for evaluating and implementing omics technologies for human and environmental safety is an essential first step. Groundtruthing is expected to take place in stages. Many parties are now engaged in the initial stage of profile or fingerprint identification. The global analytical scope of omics technologies is being used to search for a generally reproducible profile or fingerprint that can discriminate a toxicological event or mechanism. This is typically done using 2 or 3 well-understood reference test compounds. This putative profile may consist of tens to hundreds of gene transcripts, proteins, or metabolites. Multisector collaborative research initiatives, such as the one performed by the Health and Environmental Sciences Institute (HESI) Committee on Genomics in Mechanism-Based Risk Assessment, have undertaken large experimental programs to assess the feasibility of identifying consistent profiles across different laboratory settings and technical platforms (Pennie et al. 2004.). The findings indicated that although gene-by-gene expression can be highly variable, it is possible to identify commonly affected functional pathways across parties (Chu et al. 2004).

A second stage is a careful dissection of these profiles in order to identify truly reliable and biologically relevant markers. This marker subset of the profile should have at least 3 basic characteristics:

- The markers' significance to the measured effect of interest should be supported by biological plausibility, such as being up- or downregulated enzymes in the target mechanism or pathway.
- Performance reliability should generally be supported with a larger number of known positive and negative chemicals than would be needed at the first stage.
- The investigations should provide perspective on the linkage of the markers to dose or exposure, adaptation, or classical toxicological adverse effect so that they can be properly interpreted in a regulatory context. This process of achieving perspective is often termed phenotypic anchoring.

The latter would include at least some understanding of the particular limitations and variability, whether in the lab under controlled conditions or in the field with multiple stressors. These steps or stages should be similar for human and environmental applications.

An important element of any framework that is currently lacking is a clear rationale and decision framework for classifying and interpreting different markers — that is, a marker of exposure, a marker of adaptation, or a marker of adverse effect. Development of such a decision framework for accepting causality constitutes a third stage in groundtruthing. However, it is important that such development pay attention to the developing nature of omics technologies. Historically, there have been episodes when new technologies or procedures emerged and were too rapidly or too firmly incorporated into the risk assessment or management framework. They became established as absolutes that were "carved in stone" and applied to all situations, rather than being introduced on a provisional basis. A provisional basis allows case-by-case exploration and the technologies are subject to modification, improvement, or even retraction as experience should direct. Examples are the Ames assay and the use of low-dose linear extrapolation for cancer bioassays.

For omics technologies, such a rationale and decision framework might begin with criteria similar to those of Hill (1965) for epidemiology. Hill posed an analogous question to one we face: When do we move from an observed association in the omics data to the interpretation of the data and, from this, a toxicological conclusion? Many of Hill's criteria, such as the strength of the data and its association, the consistency of the findings, the specificity of a profile, the temporal sequence between transcription and phenotypical changes, the biological gradient or dose response, biological plausibility, and the subjecting of the hypothesis to experimental testing, seem promising for omics. The rapid advances in omics technologies suggest that the development of these framework criteria is an important priority that needs attention.

5.4.2.2 Implementation Issues for Omics

Currently, many regulatory organizations are working towards defining technically sound and legally meaningful definitions of genomics biomarkers for purposes of regulatory development. The USFDA's draft guidance on toxicogenomic data submissions predicates required submission in part on identification of "known valid bio-markers" via genomics analysis (FDA 2003). However, further public discussion on interpretation and implementation of the term biomarker is anticipated.

With sufficient groundtruthing, the particular omics data and systems can be well enough established for use to support regulatory decisions, but not necessarily make such decisions outright. In theory, regulators trained and familiar with omics data will be able to evaluate the submitted data and should be able to reach conclusions similar to those of the data submitter. However, experience in collaborative genomics research initiatives has demonstrated that the volume and complexity of genetic microarray data can yield multiple interpretations, depending upon the analytical approach adopted. HESI and others have organized workshops with industry and regulatory scientists to work through case-study examples of the impact of different analytical approaches. Similar training and discussion forums will continue to be valuable as the use of these data by regulatory agencies expands. The ability to verify conclusions reached in submitted data will also require sufficient bioinformatics resources to conduct appropriate levels of analyses.

Illustrative examples of potential use for omics data include screening of compounds to facilitate classification and inform future studies, supporting the interpretation of a substance's mechanism of action or of the observed NOEL and LOEL in a substance's dose response, potentially refuting spurious or equivocal data as inconsistent with the substance's mechanism of action, better understanding a drug's potential to exhibit a specific toxicity (e.g., hepatotoxicity) in humans, or facilitating hypotheses regarding level of concordance in metabolism and biotransformation of a compound between laboratory animal species and humans. However, at this time, it appears that the primary use of omics data in risk assessment is as supporting information to improve existing modes of risk assessment and risk management.

Recognizing that omics technologies are in their infancy, previous chapters describe efforts to begin the standardization of methods, what is known about the laboratory transferability and reproducibility of various methods, and the efforts under way or anticipated to further improve transferability and measure reproducibility. These chapters also describe the first efforts to use recognized reference chemicals in order to explore the potential linkage between several omics approaches and classical assessment endpoints. Thus, the first and possible second stages of groundtruthing appear to be under way. For the third stage, the efforts of the HESI collaborative omics research program have been to include current endpoints such as histopathology for comparison with omics results (see Amin et al. 2004; Waring et al. 2004).

The results to date are encouraging, but reinforce the conclusion that omics standardization, reproducibility, and relevance are still at an early stage. Looking ahead, for human risk assessment, the mechanistic or mode of action data for a number of chemicals exists in rodents and humans so that candidate reference chemicals as well as clearly identified assessment endpoints are known. Therefore, the basis for practical groundtruthing exercises exists and the testing of extrapolation across species can proceed.

What is clear is that additional efforts to complete groundtruthing of the initial systems are needed and will require collaboration of multiple laboratories as well as financial support. Such efforts may be considered test cases. These test cases need careful selection and planning to deliver successful technical advances and improvements as well as provide positive support for continued work to link omics to classical assessment endpoints. The aforementioned HESI Committee on Genomics is planning additional experimental programs in 2006 (involving industrial, academic, and government scientists in the United States, Europe, and Japan) to explore the viability of genomics to identify early markers of toxicity.

Efforts to establish more formal descriptions of test method validation have emerged in the last decade (ICCVAM 2003). Conceptually, new methods emerge from research and with experience they may become candidates for new regulatory assays or refinements of existing methods. The process of validation is to demonstrate key performance characteristics of the method so that its fitness for regulatory use, including benefits and limitations, is clear. Key characteristics include reliability and relevance.

Reliability is a measure of the degree to which a test method can be performed reproducibly within a given laboratory and among different laboratories over time. This typically requires a well-developed and understood set of procedures: ultimately, in formal validation, a protocol and standard operating procedures that can be transferred among laboratories and that yield similar or reproducible results. Relevance means the extent to which the new method (omics) can correctly predict the same positive and negative outcomes as the reference method. The outcomes may be simple preliminary hazard classification by method of action (the test compound has properties consistent or inconsistent with a potential to be a peroxisome proliferator) or attempt to predict that a chemical will elicit classical assessment endpoints (the test compound is a hepatotoxicant).

Long term, scientists and risk assessors might consider how a formal validation of an omics method might proceed or whether such a step is necessary and useful. In the interim, risk assessors need to recognize the opportunities to address and to diminish existing uncertainties as they arise. A gradual iterative process beginning with specific toxicological and regulatory questions would be appropriate rather than immediate, wholesale applications of omics technologies. This recognizes the emerging nature of the technologies and our currently limited understanding of their implications to risk management and regulatory policy. Initial steps are being taken by the regulated community to employ omics in particular situations where the opportunity to successfully inform risk assessors and risk managers with omics results is high. This process will help to reduce uncertainty around potential risk assessment applications and interpretations of omics data. Effectively, this is a graduated learning curve and an iterative exploration and improvement process. While this is dynamic, scientists and risk assessors should balance the need to move ahead on the overall learning curve of omics applications with the need to understand its limits and reduce uncertainties at a given time.

5.4.3 Institutional Limitations

5.4.3.1 Risk Assessment and Management Infrastructure Limitations

If parties do not understand how data are generated, analyzed, and interpreted, they will tend to question and even reject their use. Rationales for individual arguments or decisions based on omics data will be difficult to evaluate by decision-making authorities as abstractions further and further away from data occur (for example, as patterns are argued to be proof of disease within 1 context and are then applied to another context). Stated another way, a lack of experience with the technologies in the context of regulatory decision-making and lack of a critical mass of expertise within the decision-making authority may become practical limitations of omics use in risk assessment. Where the technologies generating the data are not internalized or owned by the authorities, they may reject the conclusions of analyses based on the data.

Similar challenges have been encountered for incorporation of other computationally intense methodologies (physiologically based pharmacokinetic models [PbPk], biologically based dose–response [BBDR] models), so this is not a unique issue for omics. However, in distinction from other data types and analysis methodologies, use of omics data will potentially involve several layers of abstraction of relatively novel data types and use of advanced and novel software and statistical techniques. Therefore, the appropriate use of omics data will be greatly facilitated if decision-makers have access to data and expertise so that the assumptions and defaults are effectively rendered transparent and rebutable within a given decision context. Furthermore, there is a risk of systematic or difficult-to-catch errors if omics is left as the sole province of a relative few experts and software packages, or if standards set in relative ignorance become dogma through institutional inertia or public perception.

As the community of practice grows and the growth of the omics knowledge base continues to accelerate, federal agencies are beginning to examine the potential implications for risk assessment and regulatory policy. One of the most critical challenges facing the regulatory risk assessment community is the limited capacity of these diverse entities to implement omics tools and evaluate data. It will be essential to train current risk assessors so that they will be prepared to understand the omics systems and methods employed to generate and statistically analyze the data, then appropriately interpret and apply genomics data in the context of a risk assessment. Risk assessors must be able to communicate the scientific underpinnings and the interpretive tools and models used to develop the risk assessment to risk managers and stakeholders alike. Along similar lines, it will be equally important to provide training to risk managers regarding the use of genomics information in risk assessments and the strengths and limitations of such data. The regulatory community should begin to collaborate on developing training modules for the interpretation and application of genomics data for risk assessments for the risk assessor and risk manager.

5.4.3.2 Phenotypic Anchoring and TSCA Liability (Safe Harbor)

Given the scientific and financial resources needed to explore omics in a collaborative manner, barriers to the participation by industry toxicology labs need to be recognized and addressed. There are concerns that omics data would be associated with mandated testing reporting requirements of the US Federal Toxic Substances Control

Act (TSCA) and FIFRA legislation. Under TSCA and FIFRA legislation, industry must report to the USEPA any previously unidentified observations from toxicology studies that could be regarded potentially as constituting an adverse health hazard to humans. Failure to report such observations within a few working days of the actual observation can result in severe civil and even criminal penalties. In the event that a regulatory agency disagrees at some point in the future, such penalties may be imposed retroactively. As a result, industry needs a considerable degree of policy certainty or a legal safe harbor in which to conduct research.

The required phenotypic anchoring experiments for omics data present a significant compliance concern. Under the TSCA and FIFRA legal requirements, industry may be required to make judgments as to whether novel genomics observations represent potential novel adverse health effects, even though any such judgments would be obviously scientifically premature. Furthermore, industry TSCA and FIFRA reports to the agency are part of the public record. Development and release of data whose relevance to human hazard and risk is not clearly understood presents substantial legal and ethical concerns, particularly for products of major commercial importance.

The interpretation by industry does not end the matter. In the event that industry decides not to report the genomics data immediately because 1) it interprets the finding as nonadverse, or 2) it believes it cannot legitimately interpret the relevance of a finding to human hazard, this can create substantial liabilities. If, at some later time, the exploratory genomics response should be judged by USEPA as truly representing an adverse response, the industry party may be subject to very large retroactive penalties to the time of the initial data observation. Such potential liabilities are then major barriers to the exploration of omics technologies. Until the USEPA clarifies how it will specifically deal with these early phenotypic anchoring research efforts (such as by creating a legal safe harbor protection for this research), industry may be reluctant to participate fully in much needed efforts to link genomics responses to traditional toxicologic responses.

5.4.3.3   Data Standards across Regulatory Agencies

In general, omics provides new data types that could be used as part of the support for many regulatory decisions and may not be unique in addressing any particular regulatory decision. The level of groundtruthing or certainty needed for the application of omics data to critical elements of decisions under particular regulatory scenarios (e.g., how certain the assignment to or exclusion from is at particular toxicity mode of action based on omics data) should be similar to or better than that for the current data types in order for the omics data to be useful to risk assessment. As such, the path to general agreement on the use of the data in regulatory risk assessment will follow that of other data types, such as biomarkers, genetic toxicity testing, and so on. Standards, criteria, and decision frameworks within and across agencies, governments, and stakeholders will be needed and thus pathways toward the development of such standards should be initiated through appropriate standard-setting processes. The same holds true for the statistical analyses of omics data. This process is likely to be facilitated (e.g., disagreements about bright line standards avoided) if methods are developed to propagate uncertainty from original measurements through biomarker sets and other uses of omics data, and then through to likelihood of disease outcome predicted or supported through the data.

The science of omics technologies is still developing and, if conventional wisdom is any indicator, these technologies will continue to advance at rapid rates into the foreseeable future. So-called laws of technology have been developed to describe the veritable "tidal wave" of information being generated (USEPA 2004). Monsanto's law states that "the amount of useful genetic information doubles every 18 to 24 months," while Dawkin's law states that the "cost of sequencing DNA base pairs halves every 27 months." In other words, the cost of microarray technology will continue to fall and the usage of the technology will become more prevalent.

This wider usage of the technology and resultant proliferation of omics information argues for a consensus approach to standards development within the scientific community to define best practices. The regulatory community at large is grappling with issues of data variability that currently limit or preclude their use in risk assessment. Variability can be found in a number of aspects, from nomenclature to choice of microarray platform to lab protocols. Common standards across regulatory agencies will move the science of risk assessment toward finding scientifically credible and defensible results. Efforts such as the minimum information about a microarray experiment (MIAME) were designed with a goal of establishing a standard for recording and reporting microarray expression data. This is an important step toward consensus and could potentially spawn the development of other databases and public repositories to share and exchange information (Brazma et al. 2001).

Finally, we encourage discussions and dialogue so that interpretation of omics technologies within the United States and international regulatory communities and even cross-national interpretation and use of omics technologies will be a proactively sought goal where efforts are exercised in advance, rather than debated due to differences arrived at by multiple stakeholders.

5.4.3.4 Privacy Act

Another potential barrier to omics technologies involves the growing concern within the research community about the implementation of the Privacy Rule in April 2003. The rule imposes new restrictions on the mining of patient health data (Kaiser 2004). The Privacy Rule stipulates that for omics data collected from humans, informed consent needs must be evaluated and met for current and future uses of the data and the biologic samples. Adequate and appropriate protection of an individual's identity must be assured before any exchange or use of the patient's health record occurs. As a result, some researchers are reluctant to propose new research based on these new requirements, citing cumbersome processes that cause delays and potentially introduce a bias into the study (i.e., patients that understand the value of research and provide their consent to share their medical information). In addition, with the increased attention on genotyping, more ambiguities are arising concerning the disposition of genetic data.

5.5 CONCLUSIONS AND RECOMMENDATIONS

Collectively, molecular and computational omics technologies hold promise as valuable tools to develop a new level of detail in toxicity data used in risk assessment to support regulatory decisions. The degree to which these technologies will be used for more predictive evaluations and stand-alone applications is not yet clear given the immature state of the science and its limited level of acceptance as applied to risk assessment. However, there is value in identifying possible applications as a means of defining a potential scientific landscape for the future and in considering current and future needs in support of or reaction to these new technologies. Because of the remaining technical and logistical challenges (also discussed here), it is important to recognize that these remain largely potential prospective benefits with regard to broad application for risk assessment. Possible future applications of omics to risk assessment could yield the following:

- improved understanding of cross-species extrapolation (laboratory to fields, model species to wildlife, rats to humans, etc.) by characterizing homologous or divergent metabolic pathways
- screening approaches resulting in identification of novel pathways of toxicity; new hypotheses that could redefine how we approach chemical hazard or risk characterization
- development of new and more sensitive biomarkers of exposure and effect for field applications
- opportunities to take advantage of datasets in humans and test animals for drug discovery and development that are relevant to cross-species extrapolation used in environmental risk assessment; pharmacogenomics has primarily as its goal to identify sources of interindividual variability in efficacy and toxicity of drugs, which may lead to greater understanding of modes of action and dose–response relationships for classes of compounds or for basic biologic processes with broader applications for risk assessment
- novel approaches to test long-standing assumptions for dose–response relationships (particularly in the low-dose range) and exposure assessment
- opportunity to characterize basic, conserved cellular and physiological processes central to toxicity
- novel approaches to such challenging issues as mixtures assessment or vulnerable life stages and susceptible-populations assessment

At the present time, however, challenges to the broad-scale application of these technologies to risk assessment and regulatory decision-making appear to be substantial. Several issues must be addressed in part or full before implementation could be adopted fully. As such, multiple initiatives are under way in the public and private sectors to discuss and develop methods for incorporating omics into risk assessment. Although these efforts may have somewhat different objectives, participants, and stakeholders, these basic recommendations should be considered by those working toward risk application of omics:

- Promote multisector discussion on current practical uses of omics technologies and realistic potential future applications for risk assessment.
- Support training of members in the regulatory community to receive and evaluate omics submissions.
- Engage in collaborative initial learning test cases where omics technologies are likely to contribute successfully to existing endpoint interpretation in meaningful and relevant ways; conduct public evaluation of studies and their results, including a discussion whether the omics data are or are not of added value for risk assessment.
- Encourage the investigation and use of diverse analytical and presentation approaches for omics data, as well as the public discussion of their impact on potential interpretation of the data in a risk assessment context.
- Resolve the current barriers to industrial participation presented by the lack of clarity as to TSCA and FIFRA reporting provisions, and consider the implementation of research-based safe harbor provisions.
- Engage key stakeholders in discussions on the scientific and regulatory viability of the use of omics as a biomarker of exposure/effect/toxicity, including challenges associated with the use of omics as a biomarker tool (e.g., technical variability, signal-to-noise ratios).
- Identify, evaluate, and address the associated ethical, legal, and societal implications of omics technologies.
- Explore opportunities to integrate risk assessments and foster collaborative efforts across disciplines.
- Solicit input from practitioners on the potential value and/or drawbacks of the validation of omics technologies in regulatory guidelines.

The repeated suggestion to consider the benefits and drawbacks of adding appropriately validated omics assays to the existing regulatory test battery for chemicals or pharmaceuticals is not unintentional. It is in the best interest of the public health to optimize the scientific accuracy and efficiency (in terms of financial and animal resources) of required testing for new compounds. Indeed, organizations such as the USEPA and the National Institute for Environmental Health Sciences' National Toxicology Program are currently involved in multisector initiatives to evaluate whether toxicity testing could be restructured to eliminate redundancy, increase accuracy and predictivity, and reduce animal usage. At the present time, omics does not appear to have a significant role in replacing any existing methodologies, but rather appears to serve primarily as a tool for discovery, preliminary screening, and hypothesis generation.

Furthermore, this chapter identified multiple infrastructure limitations within regulatory agencies that are currently barriers to the broad incorporation of omics, including the need for the following:

- Open access to appropriate data repositories. This would likely aid the community of practice; however, in many cases this access may conflict with proprietary interests or raise liability concerns. Therefore, access structures that optimize collaboration and independent inquiry while minimizing liability and loss of intellectual property rights are needed.

- Multistakeholder collaboration. This would foster transfer of knowledge among government, industry, and academia and among scientists, risk assessors, and decision makers (e.g., HESI example).
- Hiring, training, and retention of technical staff. This is needed for decision-making authorities in areas of omics, statistics, and informatics.
- Training for risk assessors, communicators, and managers who will use the data. There is first a need for them to understand what the data mean and use it when, where, and to the degree appropriate.

For omics, as with any new assay or technique, the balancing out of scientific insights versus resource requirements can be a long, seemingly stochastic route. As basic science and regulatory analysis increasingly move towards more microscopic (gene/cell/pathway)-based approaches to understanding macroscopic manifestations of effect, many of the challenges cited in this chapter will endure. However, the emerging multisector and multidisciplinary approach to these issues will provide benefits to the scientific community in terms of technical resolution and increased exchange of ideas and experience.

ACKNOWLEDGMENT

We appreciate the editorial assistance of Jana Labenia (Northwest Fisheries Science Center, Seattle, Washington) in drafting and revising this chapter.

REFERENCES

Amin RP, Vickers AE, Sistare F, Thompson KL, Roman RJ, Lawton M, Kramer J, Hamadeh HK, Collins J, Grissom S, et al. 2004. Identification of putative gene-based markers of renal toxicity. Environ Health Perspect 112:465–479.

Brazma A, Hingamp P, Quackenbush J, Sherlock G, Spellman P, Stoeckert C, Aach J, Ansorge W, Ball CA, Causton HC, et al. 2001. Minimum information about a microarray experiment (MIAME)—toward standards for microarray data. Nat Genet 29(4):365–371.

Chu TM, Deng S, Wolfinger R, Paules RS, Hamadeh HK. 2004. Cross-site comparison of gene expression data reveals high similarity. Environ Health Perspect Toxicogenomics 112:449–455.

Conolly RB, Beck BD, Goodman JI. 1999. Stimulating research to improve the scientific basis of risk assessment. Tox Sci 49:1–4.

Dellarco V. 2004. An EPA perspective: a new testing paradigm for pesticides. Abstract from SOT 2004 Annual Meeting. Toxicologist 78:S-1:254.

FDA. 2003. Draft: guidance for industry—pharmacogenomic data submissions. Draft guidance. US Food and Drug Administration. Washington, DC.

Forbes VE, Calow P. 2002. Species sensitivity distributions revisited: a critical appraisal. Human Ecol Risk Assessment 8(3):473–492.

Hansen LJ, Johnson ML. 1999. Conservation and toxicology: Integrating the disciplines. Conserv Biol 13(5):1225–1227.

Hill AB. 1965. The environment and disease: association or causation? Proc Royal Soc Med (London) 58:295–300.

[ICCVAM] Interagency Coordinating Committee on the Validation of Alternative Methods, 2003. ICCVAM guidelines for the nomination and submission of new, revised, and alternative test methods. Prepared by the ICCVAM and the National Toxicology Program (NTP) Interagency Center for the Evaluation of Alternative Toxicological Methods (NICEATM). NIH Publication No: 03-4508. Research Triangle Park, NC.

Kaiser J. 2004. Privacy rule creates bottleneck for U.S. biomedical researchers. Science 305(5681):168–169.

Lehman AJ, Fitzhugh OG. 1954. 100-Fold margin of safety. Assoc Food Drug Off USQ Bull 18:33–35.

Miracle AL, Toth GP, Lattier DL. 2003. The path from molecular indicators of exposure to describing dynamic biological systems in an aquatic organism: microarrays and the fathead minnow. Ecotoxicology 12(6):457–462.

National Research Council (NRC). 1983. Risk assessment in the federal government: managing the process. Washington DC: National Academy Press.

Pennie W, Pettit SD, Lord PG. 2004. Toxicogenomics in risk assessment: an overview of an ILSI HESI collaborative research program. Environ Health Perspect Toxicogenom 112:417–419.

Pennie WD, Tugwood JD, Oliver GJA, Kimber I. 2000. The principles and practice of toxicogenomics: Applications and opportunities. Toxicological Sciences 54(2):277–283.

Peterson CH, Rice SD, Short JW, Esler D, Bodkin JL, Ballachey BE, Irons DB. 2003. Long-term ecosystem response to the Exxon Valdez oil spill. Science 302(5653):2082–2086.

Preston BL. 2002. Indirect effects in aquatic ecotoxicology: implications for ecological risk assessment. Environ Manage 29:311–323.

Regan HM, Colyvan M, Burgman MA. 2002. A taxonomy and treatment of uncertainty for ecology and conservation biology. Ecol Appl 12(2):618–628.

Snape JR, Maund SJ, Pickford DB, Hutchinson TH. 2004. Ecotoxicogenomics: the challenge of integrating genomics into aquatic and terrestrial ecotoxicology. Aquatic Toxicol 67(2):143–154.

Stark JD, Banks JE, Vargas R. 2004. How risky is risk assessment: The role that life history strategies play in susceptibility of species to stress. Proc Natl Acad Sci U S A 101(3):732–736.

Suter G, Vermiere T, Munns W, Sekizawa J. 2001. Framework for the Integration of Health and Ecological Risk Assessment 2001. Report Prepared for the WHO/UNEP/ILO International Programme for Chemical Safety.

Timchalk C. 2004. Comparative inter-species pharmacokinetics of phenoxyacetic acid herbicides and related organic acids. Evidence that the dog is not a relevant species for evaluation of human health risk. Toxicology 200:1–19.

Touart LW, Maciorowski AF. 1997. Information needs for pesticide registration in the United States. Ecol Appl 7(4):1086–1093.

[USEPA] US Environmental Protection Agency. 1998. Guidelines for ecological risk assessment. Fed Register 63(93):26846–26924.

[USEPA] US Environmental Protection Agency. 2004. Draft: potential implications of genomics for regulatory and risk assessment applications at EPA.

Waring JF, Ulrich RG, Flint N, Morfitt D, Kalkuhl A, Staedtler F, Lawton M, Beekman JM, Suter L. 2004. Interlaboratory evaluation of rat hepatic gene expression changes induced by methapyrilene. Environ Health Perspect 112:439–448.

Wilkinson CF, Christoph GR, Julien E, Kelley JM, Kronenberg J, McCarthy J, Reiss R. 2000. Assessing the risks of exposures to multiple chemicals with a common mechanism of toxicity: how to cumulate? Reg Toxicol Pharmacol 31:30–43.

Williams TD, Gensberg K, Minchin SD, Chipman JK. 2003. A DNA expression array to detect toxic stress response in European flounder (Platichthys flesus). Aquatic Toxicol 65(2):141–157.

Zhao YA, Newman MC. 2004. Shortcomings of the laboratory-derived median lethal concentration for predicting mortality in field populations: exposure duration and latent mortality. Environ Toxicol Chem 23(9):2147–2153.

Index

Other Titles from the Society of Environmental Toxicology and Chemistry (SETAC):

Working Environment in Life-Cycle Assessment
Poulsen and Jensen, editors
2005

Life-Cycle Assessment of Metals
Dubreuil, editor
2005

Life-Cycle Management
Hunkeler, Saur, Rebitzer, Finkbeiner, Schmidt, Jensen, Stranddorf, Christiansen
2004

Scenarios in Life-Cycle Assessment
Rebitzer and Ekvall, editors
2004

Life-Cycle Assessment and SETAC: 1991–1999
(15 LCA publications on CD-ROM)
2003

Code of Life-Cycle Inventory Practice
de Beaufort-Langeveld, Bretz, van Hoof, Hischier, Jean, Tanner, Huijbregts, editors
2003

Life-Cycle Assessment in Building and Construction
Kotaji, Edwards, Shuurmans, editors
2003

Community-Level Aquatic System Studies—Interpretation Criteria (CLASSIC)
Giddings, Brock, Heger, Heimbach, Maund, Norman, Ratte, Schäfers, Streloke, editors
2002

Interconnections between Human Health and Ecological Variability
Di Giulio and Benson, editors
2002

Life-Cycle Impact Assessment: Striving towards Best Practice
Udo de Haes, Finnveden, Goedkoop, Hauschild, Hertwich, Hofstetter, Jolliet,
Klöpffer, Krewitt, Lindeijer, Müller-Wenk, Olsen, Pennington, Potting, Steen, editors
2002

Silver in the Environment: Transport, Fate, and Effects
Andren and Bober, editors
2002

Test Methods to Determine Hazards for Sparingly Soluble Metal Compounds in Soils
Fairbrother, Glazebrook, van Straalen, Tararzona, editors
2002

Avian Effects Assessment: A Framework for Contaminants Studies
Hart, Balluff, Barfknecht, Chapman, Hawkes, Joermann, Leopold, Luttik, editors
2001

SETAC

A Professional Society for Environmental Scientists and Engineers and Related Disciplines Concerned with Environmental Quality

he Society of Environmental Toxicology and Chemistry (SETAC), with offices currently in North America and Eu-
pe, is a nonprofit, professional society established to provide a forum for individuals and institutions engaged in
e study of environmental problems, management and regulation of natural resources, education, research and de-
lopment, and manufacturing and distribution.

ecific goals of the society are:

- Promote research, education, and training in the environmental sciences.
- Promote the systematic application of all relevant scientific disciplines to the evaluation of chemical hazards.
- Participate in the scientific interpretation of issues concerned with hazard assessment and risk analysis.
- Support the development of ecologically acceptable practices and principles.
- Provide a forum (meetings and publications) for communication among professionals in government, busi-
  ness, academia, and other segments of society involved in the use, protection, and management of our environ
  ment.

hese goals are pursued through the conduct of numerous activities, which include:

- Hold annual meetings with study and workshop sessions, platform and poster papers, and achievement and
  merit awards.
- Sponsor monthly and quarterly scientific journals, a newsletter, and special technical publications.
- Provide funds for education and training through the SETAC Scholarship/Fellowship Program.
- Organize and sponsor chapters to provide a forum for the presentation of scientific data and for the inter-
  change and study of information about local concerns.
- Provide advice and counsel to technical and nontechnical persons through a number of standing and ad hoc
  committees.

TAC membership currently is composed of more than 5,000 individuals from government, academia, business,
d public-interest groups with technical backgrounds in chemistry, toxicology, biology, ecology, atmospheric sci-
ces, health sciences, earth sciences, and engineering. If you have training in these or related disciplines and are
gaged in the study, use, or management of environmental resources, SETAC can fulfill your professional affiliation
eds.

ll members receive a newsletter highlighting environmental topics and SETAC activities, and reduced fees for the
nnual Meeting and SETAC special publications. All members except Students and Senior Active Members receive
onthly issues of *Environmental Toxicology and Chemistry* (*ET&C*) and *Integrated Environmental Assessment and Man-
ement* (*IEAM*), peer-reviewed journals of the Society. Student and Senior Active Members may subscribe to the
urnal. Members may hold office and, with the Emeritus Members, constitute the voting membership.

you desire further information, contact the appropriate SETAC Office.

1010 North 12th Avenue	Avenue de la Toison d'Or 67
Pensacola, Florida 32501-3367 USA	B-1060 Brussels, Belgium
T 850 469 1500 F 850 469 9778	T 32 2 772 72 81 F 32 2 770 53 83
E setac@setac.org	E setac@setaceu.org

www.setac.org

Environmental Quality Through Science®